Current Topics in Pathology
76

Neuropathology

Contributors

J.H. Adams · J.R. Anderson
C.L. Scholtz · R.O. Weller

Editor

Colin L. Berry

With 85 Figures
and 1 Colour Plate

Springer-Verlag
Berlin Heidelberg New York
London Paris Tokyo

C.L. Berry, Professor Dr., Department of Morbid Anatomy, The London Hospital Medical College, Whitechapel, London E1 1BB, Great Britain

E. Grundmann, Professor Dr., Gerhard-Domagk-Institut für Pathologie der Universität, Domagkstraße 17, D-4400 Münster

Library of Congress Cataloging-in-Publication Data. Neuropathology. (Current topics in pathology; 76) Includes bibliographies and index. 1. Nervous system – Diseases. 2. Brain damage. 3. Encephalitis. 4. Senile dementia. I. Adams, J. Hume. II. Berry, Colin Leonard, 1937– III. Series: Current topics in pathology; v. 76. [DNLM: 1. Nervous System Diseases – pathology. WI CU821H v. 76/WL 100 N494518] RB1.E6 vol. 76 [RC347] 616.07 s 87-32363 [616.8'047]

ISBN-13: 978-3-642-71355-2 e-ISBN-13: 978-3-642-71353-8

DOI: 10.1007/ 978-3-642-71353-8

Softcover reprint of the hardcover 1st edition 1988

Typesetting: Universitätsdruckerei H. Stürtz AG, D-8700 Würzburg
2122/3130-543210

List of Contributors

ADAMS, J.H., Prof.	Department of Neuropathology, University of Glasgow, Institute of Neurological Sciences, Southern General Hospital, Glasgow G51 4TF, Great Britain
ANDERSON, JANICE R., Dr.	Department of Morbid Anatomy and Histopathology, John Bonnett Clinical Laboratories, Addenbrooke's Hospital, Hill Road, Cambridge CB2 2QQ, Great Britain
SCHOLTZ, C.L., Dr.	Department of Morbid Anatomy, The London Hospital Medical College, The University of London, London E1 1BB, Great Britain
WELLER, R.O., Prof.	Department of Neuropathology, Southampton General Hospital, Faculty of Medicine, The University of Southampton, Southampton SO9 4XY, Great Britain

Preface

Increasing specialisation in pathology reflects the progressive changes in medical practise. The advent of a specialist with a new interest in a hospital or clinic may present the pathologist with a need to extend his or her knowledge to be able to work closely with the clinical practitioner in order to provide adequate clinical care.

Some sub-specialisations are long established, such a one is neuropathology. However, an exclusive specialist practise is generally confined to neurosurgical centres and much neuropathology is of necessity, executed by general pathologists.

The areas covered by this volume are those which are commonly managed by the generalist. Professor Adams' account of how the skull and brain should be examined here will give confidence to many by defining a good technique and the careful description of various kinds of vascular injury lesions resulting from raised intracranial pressure will help to clarify repeated difficulty. More subtle forms of damage are also considered in detail.

Professor Weller provides a detailed account of how the central nervous system may be examined in a way which permits all of us to prepare material which will allow adequate investigation of central nervous system disease and the proper examination of peripheral nerves. This chapter will become a "handbook" and will be of interest to those in training and established practitioners. Muscle biopsy is also dealt with; this is an area of investigative concern for many generalists. The role of that singular neuropathological technique is very clearly emphasized.

Viral encephalitis commonly presents as an acute disease in a general hospital and is dealt with in detail by Dr. Anderson. Knowledge of potential causative agents is limited in the non-specialist and the value of various modes of investigation is discussed here for arboviruses, herpes simplex, post-infectious encephalitis, sub-acute sclerosing panencephalitis and opportunistic infections including AIDS. The slow viral infections are described.

Dr Scholtz' contribution deals with a major and increasing problem in our Western ageing populations: dementia. This is an area in which the precise methodology of histopathology has an important part to play in increasing knowledge in this still understudied area. Recent developments in the examination of metals in tissues, appreciation of specific forms of vascular damage associated with dementia and the role of trauma are all considered, the role of alcohol, dialysis and other metabolic disorders in the general picture of this clinical entity are also discussed.

The topics covered here consider areas assessed by the Editors as being those where a clear account of current techniques and views will be of value to those who undertake neuropathological investigation as part of a broader pattern of work. It is hoped that they will enable the non-specialist to improve the standard of pathological support given to clinical colleagues.

London C.L. BERRY

Contents

Indexed in ISR

The Autopsy in Fatal Non-Missile Head Injuries

J.H. Adams

The identification and interpretation of brain damage resulting from a non-missile head injury is often not easy – if only because the most obvious structural damage identified *post mortem* may not be the most important. There must be many occasions when death is certified as being due to fracture of the skull and cerebral contusions when neither may have even been the reason for the patient having been unconscious. Patients with a fracture and quite severe contusions can make an uneventful and complete recovery from their injury if no other types of brain damage are present (cf. Figs. 7 and 9). The aim of this chapter therefore is to advise on a pragmatic approach to the autopsy, to define clearly the various types of brain damage that may occur, and to remind the pathologist that he must know what he is looking for. This is particularly important when there is no evidence of severe cerebral contusions or of a large intracranial haematoma. A patient may die as a result of a non-missile head injury without either of these being present. Thus a systematic approach is required, and the pathologist must be aware of the more subtle – and often most important – types of brain damage that can occur in a patient who sustains a non-missile head injury. Frequently the damage can be identified only microscopically.

1 The Autopsy

Only examination of the head will be considered here but the pathologist will be aware of the importance of looking for multiple injuries and, in particular, for evidence of injury to the cervical spine. This is a not infrequent occurrence in a patient who sustains a severe head injury in a road traffic accident, and its existence is often disclosed by the presence of haemorrhage into the paravertebral muscles in the neck.

Any damage to the scalp should be noted since this is probably the best indication of the site of any impact. Periorbital bruising and subconjunctival haemorrhage are often indicative of a fracture in the anterior cranial fossa. Once the scalp has been reflected, any fracture affecting the calvaria should be noted. Fractures are most easily recorded on line diagrams of the skull.

Ideally, the pathologist should remove the brain himself so that he can assess, for example, the tightness of the dura. He must at the very least see the brain being removed. The first important point is to ensure that enough of the calvaria is removed to make removal of the brain simple and straightforward. The saw cut in the frontal bone should be not more than 1 cm above the supra-orbital ridges; and posteriorly the cut should lie just above the external occipital protuberance. It is virtually impossible to remove a damaged and often swollen brain through an inadequate exposure. Any extradural haematoma will normally remain in position. This should be scraped off into an appropriate container and its volume measured since any haematoma less than some 30–35 ml in volume is unlikely to have acted as a fatal intracranial expanding lesion. One of the simplest ways to measure the volume of the clot is to place 100 ml of water in a broad measuring cylinder, add the clot to the cylinder, and then read the total volume. The superior sagittal sinus should then be opened to ascertain if it is patent. The dura is best opened with curved scissors along the line of the saw cut (Fig. 1), and care has to be taken to pull the dura away from the brain so that the brain is not damaged by the reverse side of the scissors. At this stage, the tightness of the dura can be assessed. An acute subdural haematoma will not, unlike an extradural haematoma, remain in a position and it must be collected in an appropriate container as the dura is being opened. Its volume should also be measured. Once the falx has been transected, an attempt should be made to identify the bridging veins, and any ruptured veins, while the dura is being reflected (Fig. 2).

Undue traction on the brain and brain stem should be avoided by using gravity as much as possible. The head, therefore, should be so positioned that the frontal poles separate spontaneously from the anterior cranial fossa. Once the olfactory bulbs have been gently separated from the base of the skull, it is often necessary to free by blunt or sharp dissection adhesions between the temporal poles and the lesser wing of the sphenoid bone. Thereafter the various structures between the brain and the base of the skull, viz. the optic nerves, the internal carotid arteries, the pituitary stalk, the remaining cranial nerves and the vertebral arteries just as they enter the skull are transected. Curved scissors (Fig. 3) are again preferable since they do not tear the structures being cut. During removal of the brain, the pathologist should ascertain that no

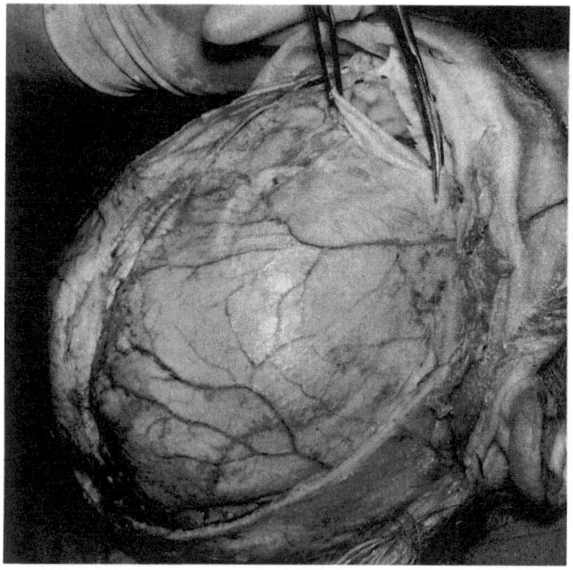

Fig. 1. The dura is incised along the line of the saw cut with with curved scissors. The meningeal arteries are clearly seen on the surface of the dura. (From ADAMS and MURRAY 1982)

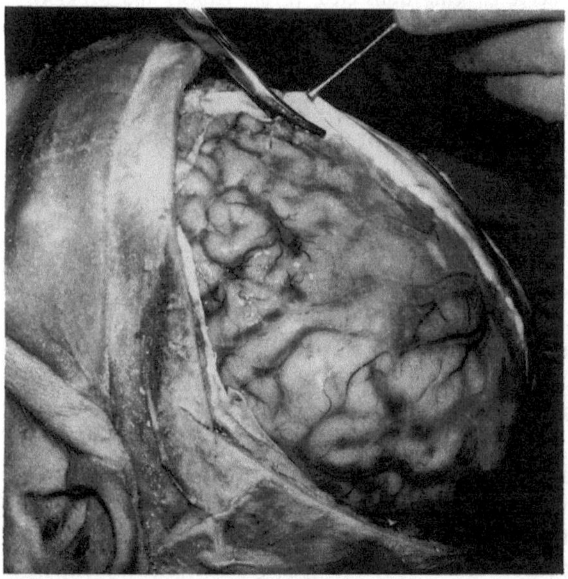

Fig. 2. Once the dura has been reflected, the bridging veins are identified and transected. (From ADAMS and MURRAY 1982)

structures, for example the pituitary stalk, have been torn across at the time of injury. The final stage is to transect the upper cervical cord transversely before gently delivering the brain. A deep wedge-shaped incision should not be made since this interferes with proper examination of the upper cervical segments if damage of the spinal cord is suspected.

The next stage is to strip the dura from the base of the skull to reveal any fractures, which are again best recorded diagrammatically. Pressure should

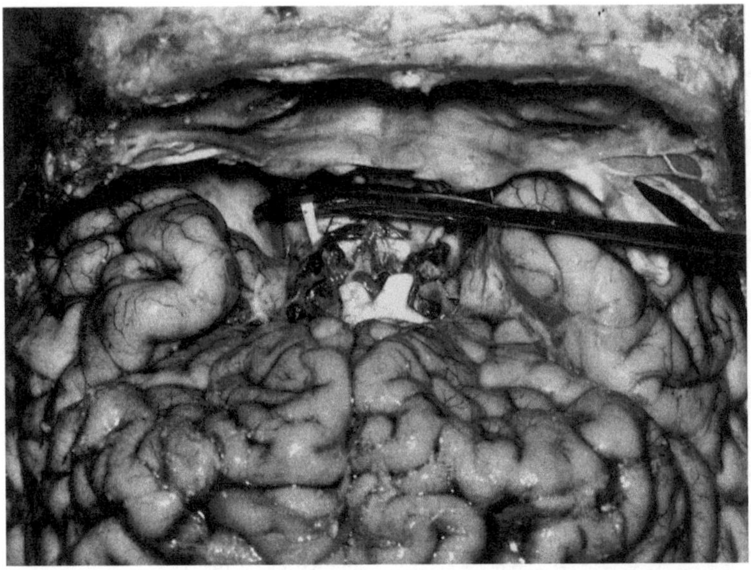

Fig. 3. Removal of the brain. All the structures between the brain and the base of the skull must be cut individually. Here an oculomotor nerve is being transected. (From ADAMS and MURRAY 1982)

be applied to the frontal and occipital bones to establish whether or not a 'hinge' fracture across the base of the skull is present. Hairline fractures of the calvaria are often seen more clearly if it is filled with water.

1.1 Fracture of the Skull

The presence of a fracture usually means that impact has been severe, but in about 20–25 per cent of fatal non-missile head injuries there is no fracture. Impact with a blunt object usually produces a *linear* fracture, and several fracture lines may radiate outwards from the point of impact. Impact with a sharp object characteristically produces a *depressed* fracture. Fractures of the base of the skull are commonly encountered in severe head injuries, while a *'hinge'* fracture extending right across the base of the skull, usually in the region of the clivus, indicates that the injury has been particularly severe. With injuries to the occipital region, pressure waves through the brain may cause *contrecoup* fractures of the orbital plates. Two important complications of a fracture of the skull are extradural haematoma and infection.

1.2 Examination of the Brain

Even when the cause of death is obvious, e.g. when there is large extradural or subdural haematoma and evidence of tentorial herniation, the brain should be fixed intact prior to dissection. This becomes essential if there are no obvious

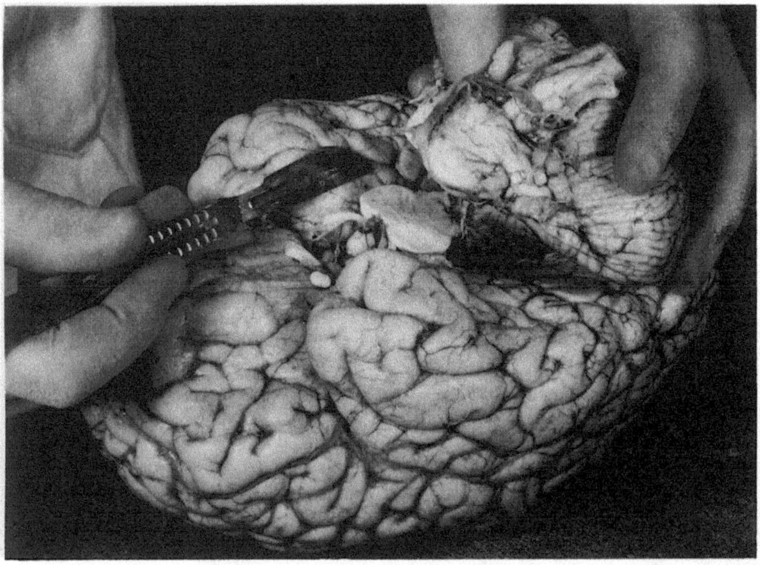

Fig. 4. Dissection of the brain. The first cut is normally made through the rostral pons or the caudal midbrain. (From ADAMS and MURRAY 1982)

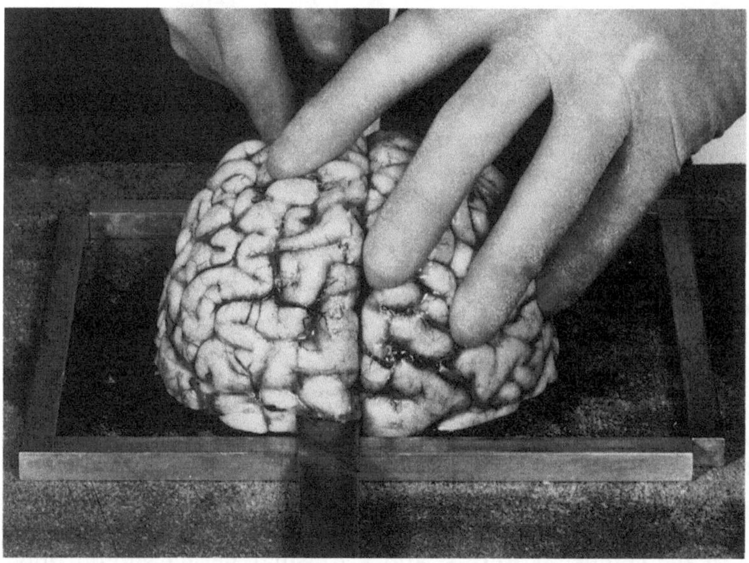

Fig. 5. Dissection of the brain. With the help of cutting angles, the brain can be sliced into sections of equal thickness, normally 1 cm. (From ADAMS and MURRAY 1982)

external abnormalities. Fixation is most easily done by suspending the brain from the basilar artery in 10 per cent formol saline. The formalin should be changed after 3 days and then at weekly intervals. Care must be taken not to damage the basilar artery when the formalin is being changed. The brain should be fixed for between 3 and 4 weeks.

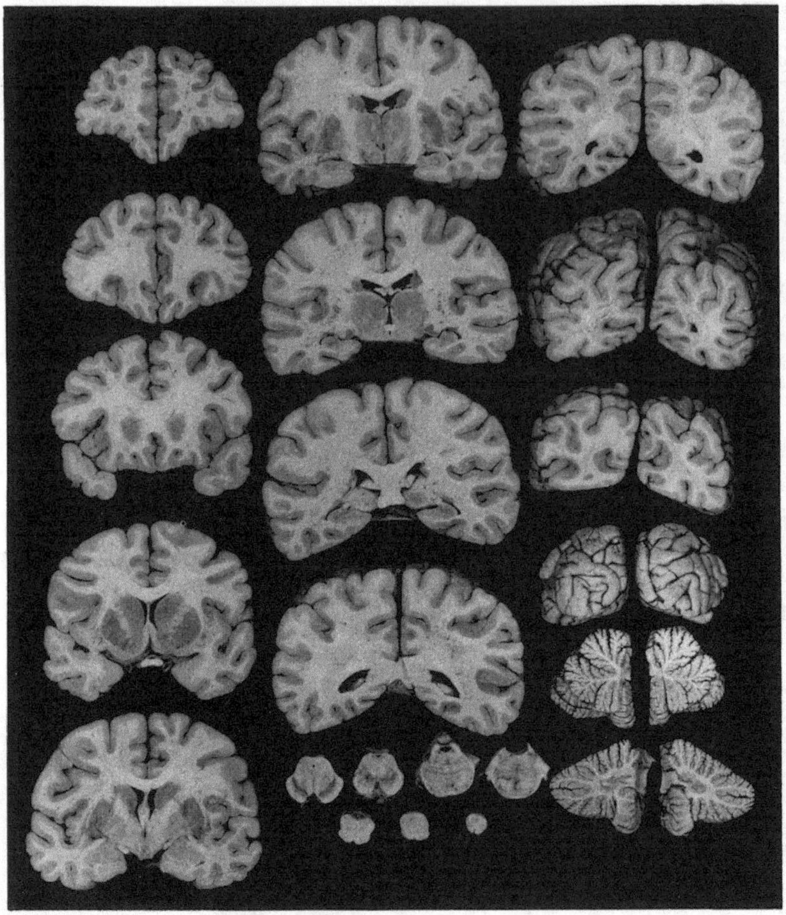

Fig. 6. Dissection of the brain. The brain laid out after being dissected as described in the text. (From ADAMS and MURRAY 1982)

The brain should first be assessed for external abnormalities such as contusions (which again are most easily recorded diagrammatically), flattening of the convolutions and internal herniae. Tentorial herniation is best seen once the midbrain has been transected (Fig. 4). There is great variation in the anatomical configuration of the cerebellar tonsils, and a tonsillar hernia should only be said to be present if the tips of the tonsils are haemorrhagic and/or there is a tranverse depression on the ventral surface of the medulla where it has been compressed against the anterior margin of the foramen magnum. A major problem that pathologists now encounter is the so-called *'respirator' brain*: once brain death has occurred, the brain swells, there is loss of definition of intrinsic anatomical markings and the cerebellar tonsils obstruct the foramen magnum and then disintegrate with the result that cerebellar tissue can be found as far down the vertebral canal as the lumbar region.

After external examination, the brain should be dissected in a standard fashion (ADAMS and MURRAY 1982); in most circumstances the most informative technique is to slice the cerebral hemispheres in the coronal plane, the cerebellum (once it has been separated from the brain stem) at right angles to the folia, and the brain stem horizontally. The use of cutting angles (Fig. 5) allows the cerebral hemispheres to be cut at a standard thickness, usually 1 cm.

The slices of brain can then be assessed for all of the various types of brain damage described below (Fig. 6). Once again, diagrammatic records on a series of line diagrams of the cerebral hemispheres, the cerebellum and the brain stem may be more informative than a written description. Particular attention should be paid to the depth and distribution of contusions; the presence of gliding contusions, intracerebral hematoma and primary damage in the corpus callosum and in the brain stem; and evidence of a high intracranial pressure as shown by displacement of the midline structures, internal herniae and secondary haemorrhage in the brain stem.

Blocks to be taken for histological examination will depend to a considerable extent on the type of brain damage suspected to be present as a result of macroscopic examination of the brain.

2 Brain Damage in Non-Missile Head Injury

There has in the past been a tendency to classify the brain damage as being *primary*, i.e. occurring at the moment of injury, and *secondary*, i.e. a complication of the original injury (ADAMS et al. 1982a). This has much to recommend it when the pathologist is trying to define the sequence of events leading to the fatal outcome. There is, however, an increasing tendency to think of the brain damage as being *focal* or *diffuse* (ADAMS and GRAHAM 1984). The principal types of focal brain damage are contusions, intracranial haematoma and damage secondary to a high intracranial pressure, shift and distortion of the brain, and in the current era of sophisticated non-invasive scanning techniques the clinician may well know that these types of brain damage are present. The principal types of diffuse brain damage are diffuse axonal injury, hypoxic damage and diffuse brain swelling, and although the clinician may suspect that diffuse brain damage is present – usually because the patient is unconscious in the absence of any severe focal brain damage and a high intracranial pressure – its nature is not easy to define precisely during life.

2.1 Contusions and Lacerations

The distinction between these two types of brain damage is somewhat artificial since the term contusion is used when the pia is not breached and the term laceration when there is disruption of the surface of the brain. Their distribution, however is essentially the same. They may occur at the point of impact, particularly if there is a depressed fracture: they are then referred to as *coup* contusions.

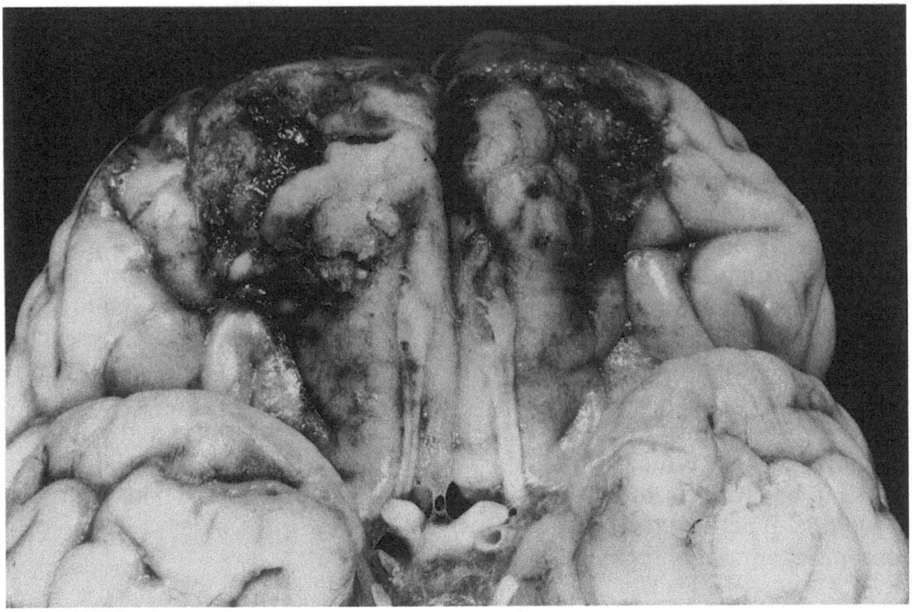

Fig. 7. Contusions. There are acute haemorrhagic contusions affecting the orbital surfaces of the frontal lobes

They are, however, basically caused by movement of the brain within the skull at the time of injury and, despite the site of any impact, occur principally where the surface of the brain comes in contact with bony protuberances. They therefore occur particularly at the frontal poles, on the undersurfaces of the frontal lobes, at the temporal poles and on the undersurfaces of the temporal lobes (Fig. 7). They also tend to occur above and below the Sylvian fissures as a result of impact against the lesser wings of the sphenoid bone, but these contusions can only be assessed in coronal sections of the brain (Fig. 8). In the early stages, contusions are haemorrhagic and are more severe at the crests of gyri than within sulci, but they often extend into the subcortical white matter. With the passage of time they become shrunken but retain a brown discoloration because of the persistence of haemosiderin. It is remarkable how frequently one finds old cerebral contusions in an apparently normal individual as an incidental finding post mortem, and it can be said quite categorically that the patient must have sustained a head injury some time in the past (Fig. 9). It is for this reason that a patient cannot really have been said to have died as a result of cerebral contusions unless there is some other complicating factor. Although contusions have long been considered the hallmark of a non-missile head injury, a patient may die as a result of such an injury without there being any contusions or lacerations of the brain.

As indicated above, the term *coup contusion* is often used for lesions that occur at a known point of impact. The term *contrecoup contusion* has frequently been used to describe contusions diametrically opposite to the point of impact,

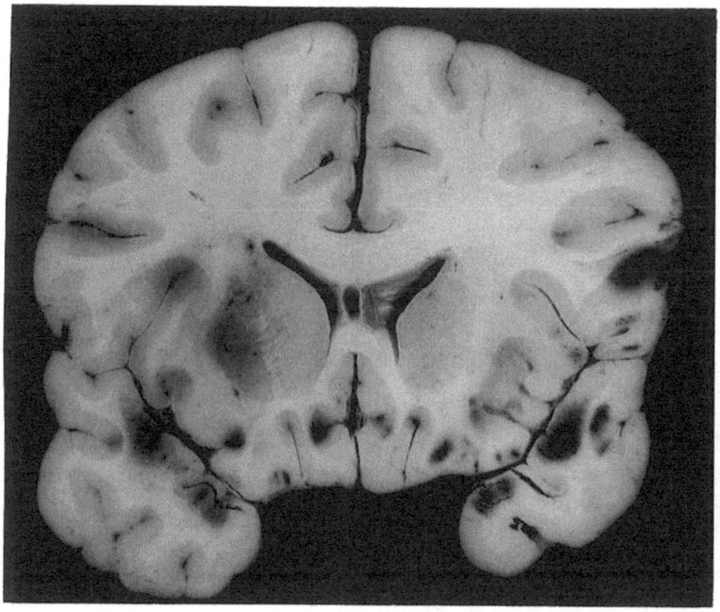

Fig. 8. Contusions. There are conspicuous haemorrhagic contusions in the cortex above and below the Sylvian fissures

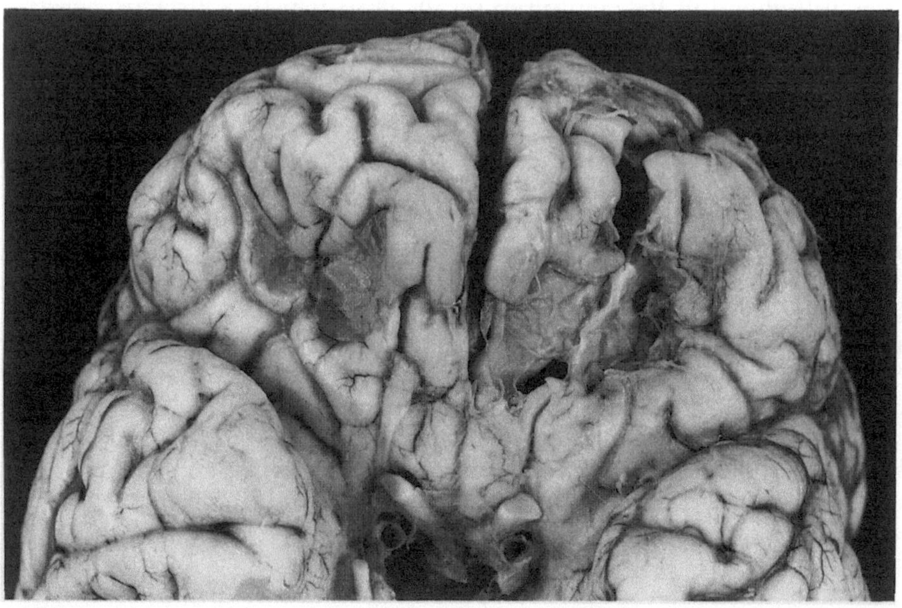

Fig. 9. Contusions. There are old shrunken contusions affecting the orbital surfaces of the frontal lobes. These were an incidental finding *post mortem* in a patient with no clinical history of a head injury. (Cf. Fig. 7)

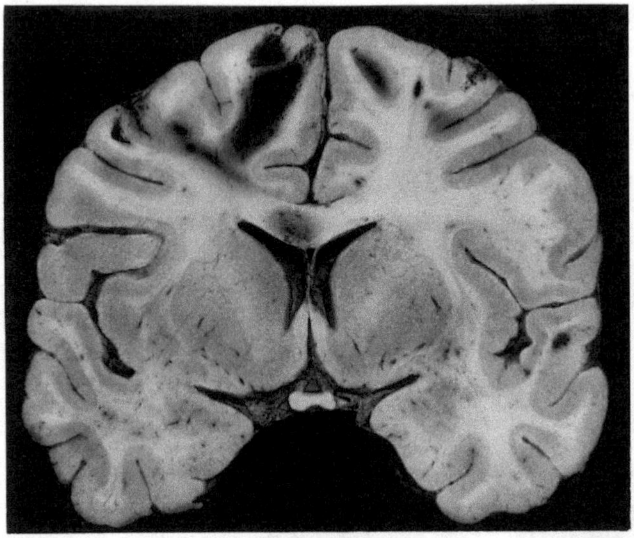

Fig. 10. Contusions. There are gliding contusions in the parasagittal white matter. There is also a haemorrhagic lesion in the corpus callosum of the type seen in diffuse axonal injury. (See Fig. 14 et seq.)

and it has been suggested that contrecoup contusions are always more severe than coup contusions. There is therefore a tendency to define the site of impact from the distribution of the contusions on the basis that they are most severe at the point diametrically opposite the point of impact. With occipital injuries, contusions are more severe in the frontal region; but with frontal injuries they are also more severe in the frontal region. Our own studies (ADAMS et al. 1985) have failed to confirm that contusions are regularly more severe diametrically opposite to the point of impact, and we feel that too much emphasis has been placed on this concept. It has to be accepted, however, that with injuries in the temporal region, contusions are often more severe in the contralateral temporal lobe.

The term *gliding contusions* is used to describe haemorrhagic lesions in the parasagittal regions (LINDENBERG and FREYTAG 1960). They may be restricted to the cortex but, more often, they extend into the adjacent white matter (Fig. 10). Sometimes they take the form of narrow lozenge-shaped haematomas within the white matter. They are related to a particular acceleration-deceleration profile at the time of injury and occur more commonly in road traffic accidents than in other types of head injury. In patients with gliding contusions there is an increased incidence of deep intracerebral haematomas and diffuse axonal injury (ADAMS et al. 1986a).

2.2 Intracranial Haematoma

These may be extradural or intradural.

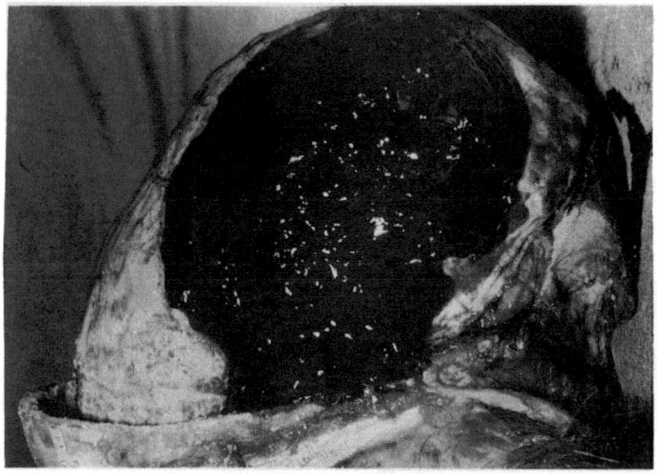

Fig. 11. Extradural haematoma. The haematoma is sharply circumscribed and usually remains attached to the dura when the calvaria is removed

2.2.1 Extradural Haematoma

This is a complication of a fracture of the skull although extradural haematoma may occur in young children in the absence of a fracture. Extradural haematomas occur most commonly in the temporal region as a result of a fracture disrupting the middle meningeal artery. The subsequent haemorrhage progressively strips the dura from the skull (Fig. 11) and, as the haematoma increases in size, it produces distortion and herniation of the brain as well as the various types of brain damage secondary to a high intracranial pressure described below. Extradural haematoma may occasionally occur in other sites and, sometimes, is bilateral.

2.2.2 Intradural Haematoma

These take four forms – 'pure' subdural haematoma, intracerebral haematoma, a 'burst' lobe and deep intracerebral (basal ganglia) haematoma.

A *pure subdural haematoma* is brought about by disruption of the parasagittal bridging veins at the time of injury. Unlike an extradural haematoma, which tends to be relatively circumscribed because of the attachment of the dura to the skull, a subdural haematoma spreads widely throughout the subdural space, not only on the lateral surface of the hemisphere, but also between and beneath the hemispheres. Any significant effects are usually due to the haematoma acting as an acute intracranial expanding lesion but, even with a small haematoma, there may be acute swelling of the underlying cerebral hemisphere (see below). A particularly characteristic appearance of the brain is the absence of flattening of the surface of the ipsilateral hemisphere because of the presence of the haematoma, and of flattening of the contralateral hemisphere which is displaced against the dura.

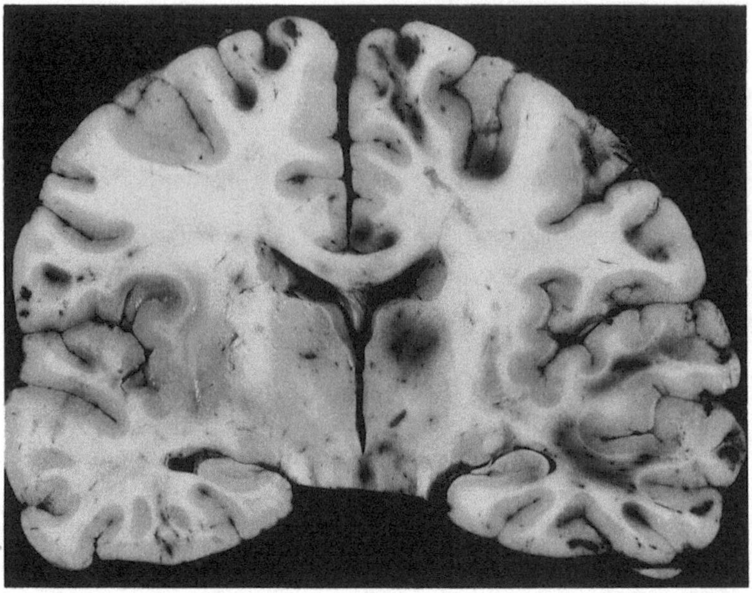

Fig. 12. Deep intracerebral haematoma. There is a haematoma in the right thalamus. There are also conventional surface contusions affecting the right temporal lobe, and bilateral gliding contusions

Intracerebral haematomas occur in association with contusions and therefore occur particularly in the frontal and in the temporal lobes.

The term *burst lobe* is used for the combination of an intracerebral haematoma, contusions and a subdural haematoma. It is almost always indicative of particularly severe brain damage.

It is only relatively recently, mainly as a result of CT scanning, that greater attention has been paid to deep intracerebral (basal ganglia) haematomas (ADAMS et al. 1986b). These haematomas occur in the striatum and the thalamus (Fig. 12) and they are related more to diffuse than focal brain damage since they have an increased incidence in patients with diffuse axonal injury.

2.3 Brain Damage Resulting from a High Intracranial Pressure

It should be clear from the foregoing that an acute intracranial expanding lesion is a common occurrence in patients who die as a result of a non-missile head injury. As the haematoma expands, the following sequence of events occurs. The surface of the brain becomes flattened and dry because of displacement of cerebrospinal fluid: there is a shift of the midline structures, the ipsilateral ventricle becomes small, the cingulate gyrus herniates under the falx (a supracallosal hernia), the parahippocampal gyrus herniates through the tentorial incisura (a tentorial hernia) and there is shift and compression of the midbrain (Fig. 13). Thus, a common terminal event in patients with an intracranial haematoma resulting from a non-missile head injury is secondary haemorrhage and

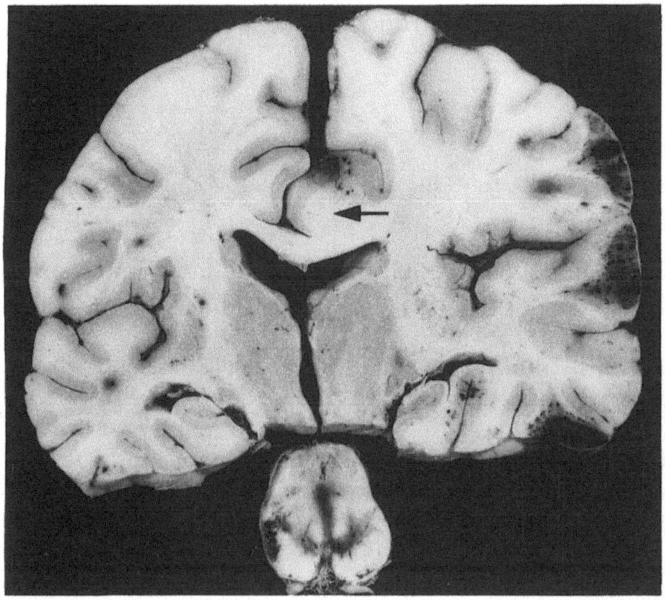

Fig. 13. Raised intracranial pressure. There are contusions on the lateral surface of the right cerebral hemisphere that were deep to an acute subdural haematoma. There is severe herniation of the cingulate gyrus (*arrow*; a supracallosal hernia), displacement of the midline structures to the left and distortion of the ventricular system. There is also secondary haemorrhage into the midbrain

infarction of the brain stem. Other features that the pathologist may note *post mortem* are focal haemorrhage in the oculomotor nerve related to a tentorial hernia, haemorrhagic infarction in the gobus pallidus or the thalamus due to compression of the perforating arteries, and infarction in the medial occipital cortex due to compression of a posterior cerebral artery.

2.4 Other Types of Focal Brain Damage

These are relatively uncommon but may be a major factor accounting for the fatal outcome. They include avulsion at the ponto-medullary junction, transection of the pituitary stalk leading to infarction of the pituitary gland, or the avulsion of various cranial nerves. Relatively recently attention has been drawn to tearing of a vertebral artery, particularly in association with blows adjacent to the angle of the mandible, where a bruise is often present. The precise pathogenesis is obscure but it would appear, sometimes in association with a fracture of the atlas, that the vertebral artery is disrupted and there is subsequent massive subarachnoid haemorrhage into the posterior fossa. This rapidly leads to tonsillar herniation, medullary compression, apnoea and death. A further complication that may result from a blow on the site of the neck is a dissecting aneurysm of the internal carotid artery which is subsequently followed by thrombosis of the artery and cerebral infarction. The carotid arteries in the neck should therefore be examined routinely post mortem.

2.5 Diffuse Brain Damage

2.5.1 Diffuse Axonal Injury

There is increasing evidence that diffuse axonal injury is the most important single factor governing the outcome in non-missile head injury (Adams et al. 1977; Adams et al. 1982b; Gennarelli et al. 1982; McLellan et al. 1986). It is the commonest cause of post-traumatic coma in the absence of an intracranial expanding lesion, and of the vegetative state and severe disability after head injury. It occurs particularly in patients who have been involved in road traffic accidents; there is an increased incidence of gliding contusions and of deep intracerebral haematomas; and there is a low incidence of intracranial haematoma and of evidence of a high intracranial pressure. Conventional surface contusions are often minimal. There have, as yet, been no reports of diffuse axonal injury occurring as a result of a fall unless this has been from a considerable height. Pathologists must therefore be acutely aware of the structural abnormalities that occur in this type of brain damage.

There are three particular features of diffuse axonal injury – focal lesions in the corpus callosum and in the dorsolateral quadrant or quadrants of the rostral brain stem, and diffuse damage to axons. The lesions in the corpus callosum (Fig. 14) and in the dorsolateral quadrant of the brain stem (Fig. 15) can usually be identified macroscopically since they are often haemorrhagic. The lesions in the corpus callosum usually lie lateral to the midline, affect the inferior part of the corpus callosum, and extend over several centimetres. They are often particularly conspicuous at the level of the splenium. They frequently extend to the midline when there is also disruption of the interventricular

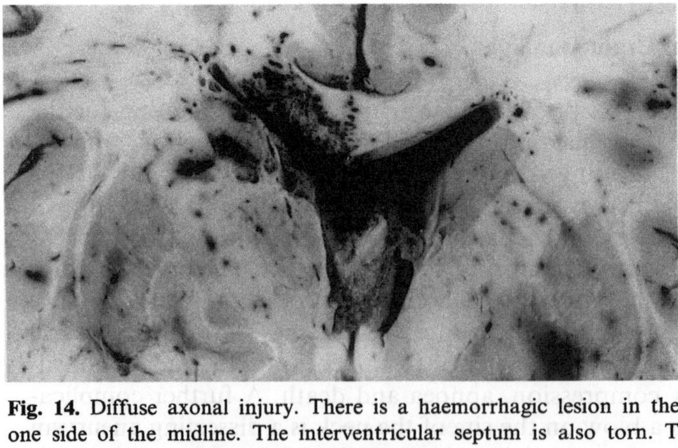

Fig. 14. Diffuse axonal injury. There is a haemorrhagic lesion in the corpus callosum mainly to one side of the midline. The interventricular septum is also torn. There are small haematomas in the caudate nucleus and in the globus pallidus

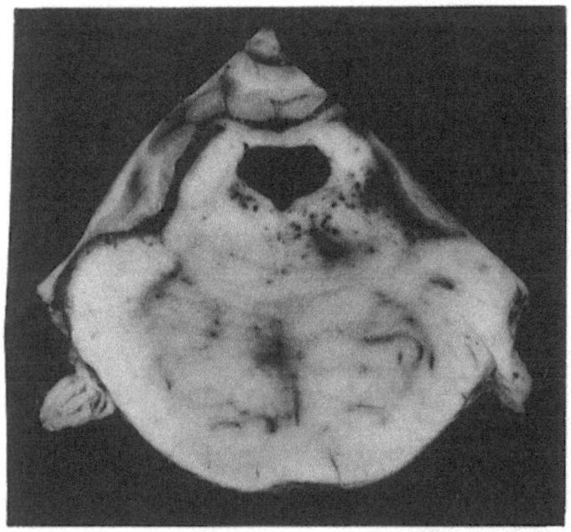

Fig. 15. Diffuse axonal injury. There is a haemorrhagic lesion in the dorsolateral quadrant of the rostral pons

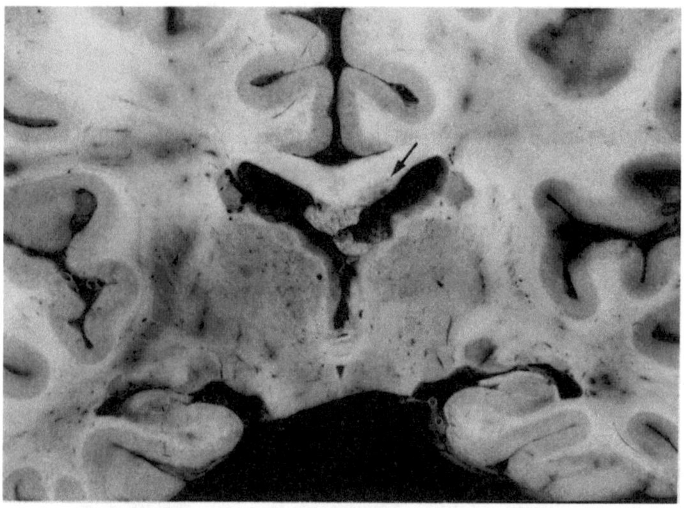

Fig. 16. Diffuse axonal injury. After a few weeks the lesion in the corpus callosum (*arrow*) becomes rather granular and more difficult to identify

septum, leading to some intraventricular haemorrhage. With the passage of time, these focal lesions come to be represented by relatively inconspicuous rather granular lesions that may be difficult to recognise unless sought specifically (Fig. 16). On histological examination there is rarefaction of tissue in the focal lesions in addition to haemorrhage in the early stages of their evolution. Within a few days there are reactive changes in astrocytes and microglia and, if the patient survives in a vegetative state for weeks or months, the focal lesions are ultimately shrunken and often cystic.

The diffuse injury to axons can only be seen microscopically and occurs in all regions of the brain but particularly in the parasagittal white matter, in the corpus callosum, and in certain tracts in the brain stem, especially the

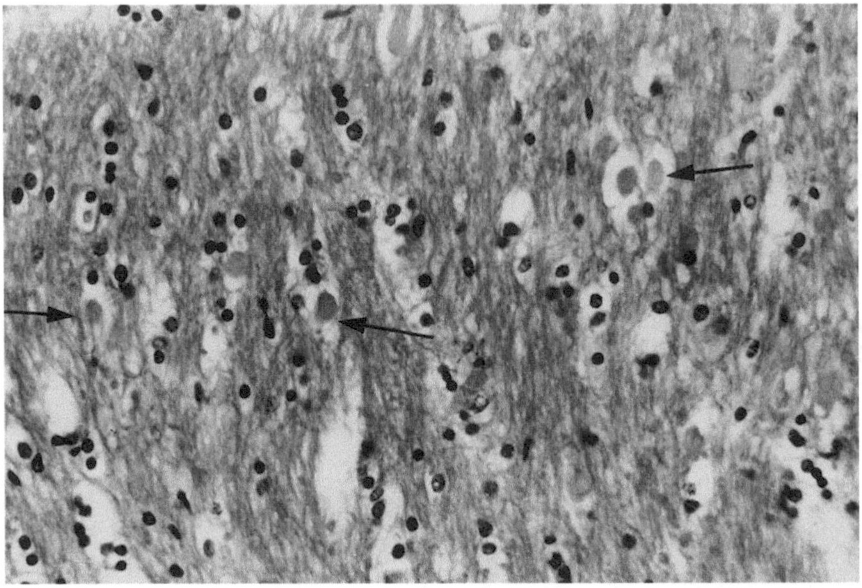

Fig. 17. There are numerous axonal retraction balls (*arrows*) in this rather coarsely vacuolated area in the corpus callosum. H & E, × 250

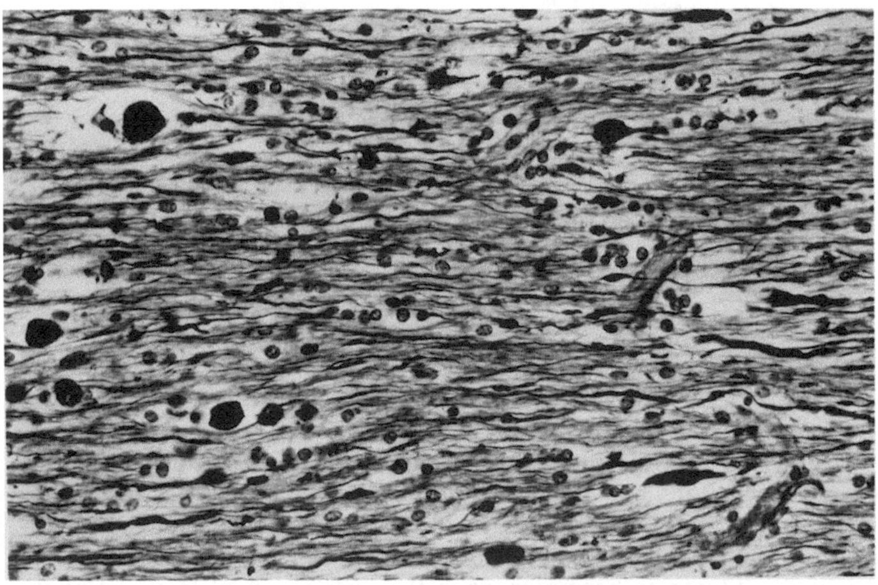

Fig. 18. Diffuse axonal injury. Axonal retraction balls are more easily identified in silver impregnation preparations. (Cf. Fig. 17). Palmgren, × 250

medial lemnisci and the corticospinal tracts in the midbrain and pons. Within days of the injury, the most conspicuous finding is the presence of axonal retraction balls in these regions. They appear as eosinophilic balls in sections stained by H & E (Fig. 17), but they are more easily identifiable in sections

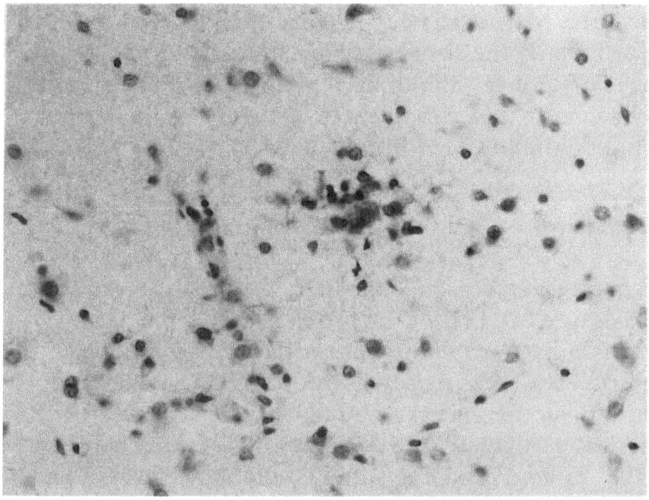

Fig. 19. Diffuse axonal injury. There is a cluster of microglia adjacent to a small blood vessel in the white matter. Cresyl violet (20 μ section), ×250

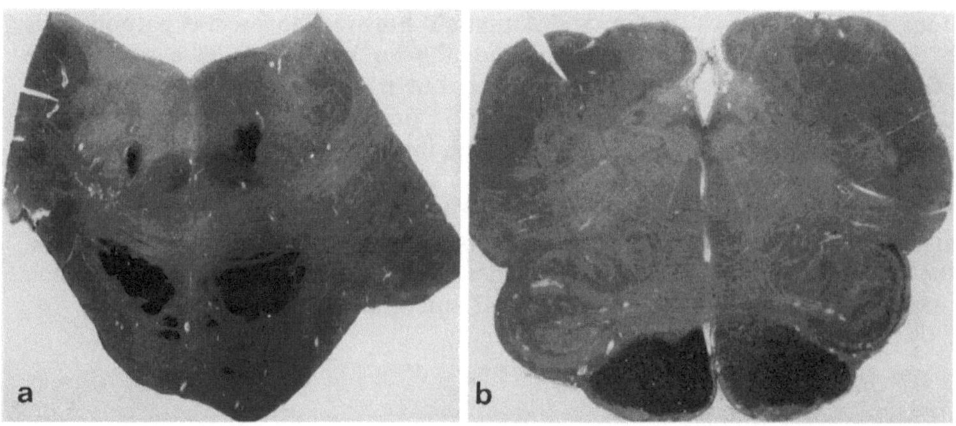

Fig. 20a, b. Diffuse axonal injury. There is degeneration of long tracts in **a** the pons and **b** the medulla in a patient who survived vegetative for 9 months after his injury. Marchi preparations

stained by silver techniques (Fig. 18). After a few weeks, the most conspicuous feature is the presence of multiple small clusters of microglia in the white matter which are best seen in thick sections (20 μ) stained by cresyl violet (Fig. 19). Some patients with diffuse axonal injury may survive for many months or even years in a vegetative state and, *post mortem*, the most striking feature is Wallerian-type degeneration in long tracts (Fig. 20) secondary to the primary damage to axons. By this time there is ventricular enlargement secondary to the loss of bulk of the white matter.

Thus, in fatal non-missile head injury where the more easily recognised types of brain damage viz. intracranial haematoma, cerebral contusions and evidence of a high intracranial pressure are absent, it must be in the forefront of the pathologist's mind that the patient has sustained diffuse axonal injury. Hence the need to examine particularly the corpus callosum and the dorsolateral quadrants of the brain stem, and to screen the brain histologically for evidence of diffuse axonal injury. The regions that should be examined are the corpus callosum, the parasagittal white matter and the brain stem.

2.5.2 Hypoxic Brain Damage

This is seen frequently, in varying degrees, in patients who die as a result of a non-missile head injury (GRAHAM et al. 1978). The problem facing the pathologist is that severe hypoxic brain damage may not be identifiable macroscopically for several days after the injury, even in a properly fixed brain. Hypoxic brain damage is particularly common in the arterial boundary zones between the major cerebral arterial territories, particularly in the parasagittal cortex between the territories supplied by the anterior and the middle cerebral arteries. The second most common type of damage is diffuse hypoxic brain damage that is not related to any particular arterial regions but, as in any type of diffuse hypoxic brain damage, it tends to increase in intensity from the frontal and the temporal poles to the occipital lobes. It also tends to be particularly conspicuous in the thalamus and in the Ammon's horns. The precise pathogenesis of hypoxic brain damage in an individual who has sustained a non-missile head injury has not yet been established but it appears to occur soon after the injury and to be related to a low blood pressure, poor oxygen saturation related to airways obstruction and to a high intracranial pressure. It is a further cause of severe disability and the vegetative state after head injury (Fig. 21).

Thus, the pathologist must screen the brain for hypoxic brain damage by at least examining selected areas of the brain, such as the parasagittal cortex, the thalamus, the Ammon's horns and the cerebellum. Failure to do so may mean that the principal cause of the fatal outcome is not established.

2.5.3 Brain Swelling

This is another poorly understood occurrence and the term 'swelling' is used rather than oedema since expansion of the white matter can occur so soon after head injury that it is attributed to congestive brain swelling rather than to true cerebral oedema resulting from an increased water content of brain tissue.

There is almost invariably some local vasogenic oedema in relation to contusions, and the term diffuse brain swelling should be restricted to diffuse swelling of one or both cerebral hemispheres. There is a close association between diffuse swelling of one cerebral hemisphere and an ipsilateral subdural haematoma. When the haematoma is evacuated, the brain simply expands to fill the space previously occupied by the haematoma (Fig. 22). Thus, the patient dies as a result of an acute intracranial expanding lesion. The swelling may not be obvious if the brain is sectioned prior to fixation. Diffuse swelling of both cerebral hemispheres tends to be restricted to children and adolescents

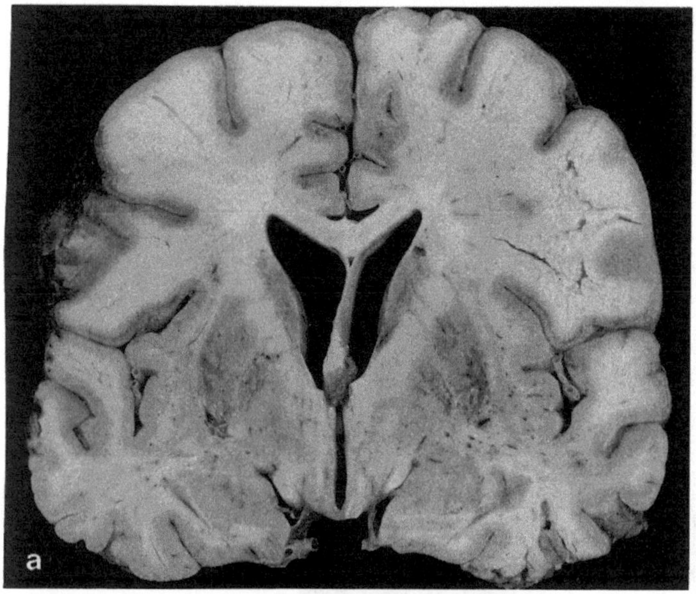

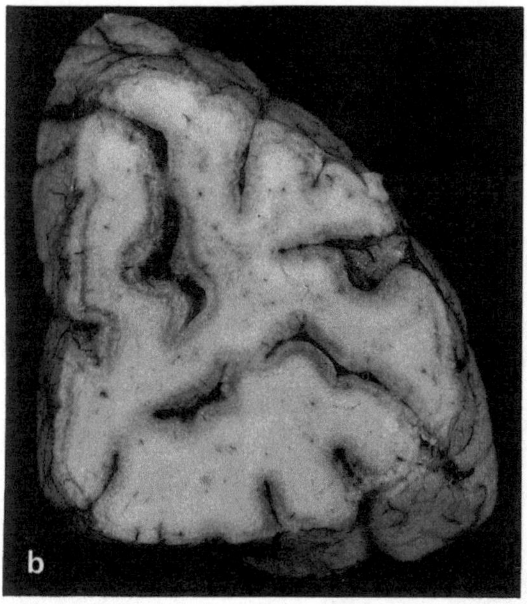

Fig. 21 a, b. Hypoxic brain damage. There is severe hypoxic damage in the brain of this patient who survived for 4 weeks after a head injury. Note **a** the disintegration of the white matter and the basal nuclei, and **b** the thin granular cortex in the occipital lobe

(Fig. 23). For some reason there is diffuse expansion of both cerebral hemispheres leading to a fatal increase in intracranial pressure. Sometimes the initial injury may have been severe, but on other occasions it may appear to have been trivial, clinical deterioration being delayed for some hours after the original injury. The clinician may think that an intracranial haematoma has developed but CT scanning simply shows diffuse swelling of the brain. The pathogenesis

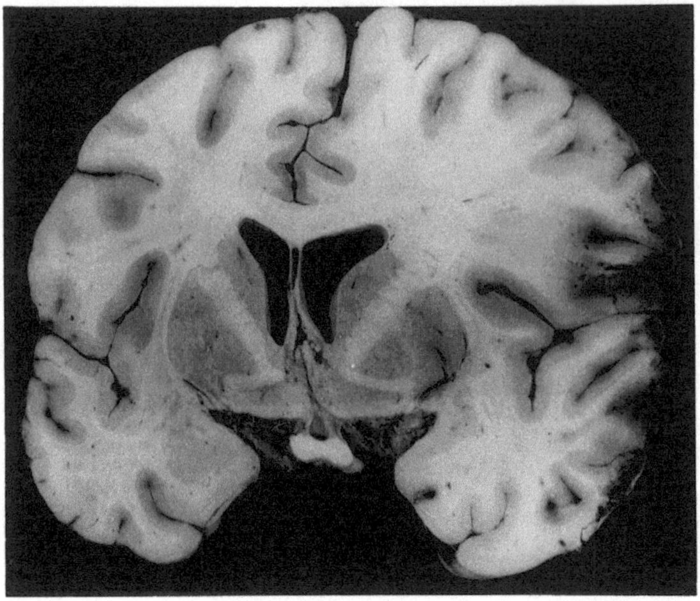

Fig. 22. Brain swelling. There is diffuse swelling of the right cerebral hemisphere. An ipsilateral acute subdural haematoma had been evacuated 4 days earlier

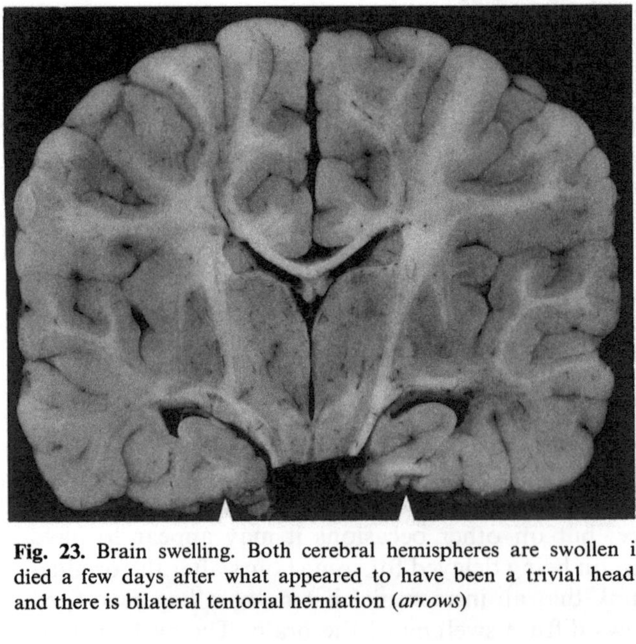

Fig. 23. Brain swelling. Both cerebral hemispheres are swollen in this 8 month old infant who died a few days after what appeared to have been a trivial head injury. The ventricles are small and there is bilateral tentorial herniation (*arrows*)

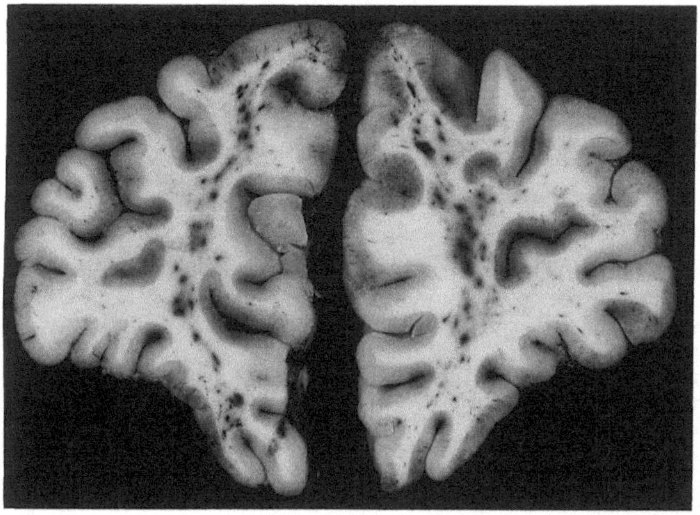

Fig. 24. Multiple patechial haemorrhages. This patient was reported to have died within minutes of a head injury

of the swelling is not understood but it seems likely that it is attributable to vasomotor paralysis leading to vasodilatation and, ultimately, to vasogenic oedema.

2.5.4 Multiple Petechial Haemorrhages

In patients who die very soon after head injury there may be multiple petechial haemorrhages in the brain (Fig. 24). The pathogenesis of this type of brain damage is also not known, and it can only be assumed that the acceleration/deceleration profile at the time of injury leads to the disruption of small blood vessels throughout the brain.

3 Conclusion

It has been the aim of this chapter to make the pathologist aware of the problem of the interpretation of brain damage in fatal non-missile head injury. The situation is clear if there is a large intracranial haematoma and evidence of brain damage secondary to a high intracranial pressure. But even in these cases there is the possibility of the co-existence of the more subtle types of diffuse damage. If, however, focal brain damage is minimal or even absent, the pathologist must pause for thought. Such a brain should always be fixed prior to dissection so that an accurate assessment of the presence of diffuse axonal injury, hypoxic brain damage or brain swelling can be made.

Acknowledgement. It is a pleasure to thank the Department of Medical Illustration, particularly Mrs. CAROLINE EADIE, of the Southern General Hospital for their help with the illustrations. Mrs. MARGARET MURRAY kindly provided Figs. 1 to 6.

References

Adams JH, Mitchell DE, Graham DI, Doyle D (1977) Diffuse brain damage of immediate impact type. Brain 100:489–502

Adams JH, Murray MF (1982) Atlas of post mortem techniques in neuropathology. Cambridge University Press, Cambridge

Adams JH, Gennarelli TA, Graham DI (1982a) Brain damage in nonmissile head injury: observations in man and subhuman primates. In: Smith WT, Cavanagh JB (eds) Recent advances in neuropathology, II. Churchill Livingstone, London, p 165

Adams JH, Graham DI, Murray LS, Scott G (1982b) Diffuse axonal injury due to non-missile head injury in humans: an analysis of 45 cases. Ann Neurol 12:557–563

Adams JH, Graham DI (1984) Diffuse brain damage in non-missile head injury. In: Anthony PP, MacSween RNM (eds) Recent advances in histopathology, 12. Churchill Livingstone, Edinburgh, p 241

Adams JH, Doyle D, Graham DI, Lawrence AE, McLellan DR, Gennarelli TA, Pastuszko M, Sakamoto T (1985) The contusion index: a reappraisal in human and experimental non-missile head injury. Neuropathol Appl Neurobiol 11:299–308

Adams JH, Doyle D, Graham DI, Lawrence AE, McLellan DR (1986a) Gliding contusions in non-missile head injury in humans. Arch Pathol Lab Med 110:485–488

Adams JH, Doyle D, Graham DI, Lawrence AE, McLellan DR (1986b) Deep intracerebral ('basal ganglia') haematomas in fatal nonmissile head injury in man. J Neurol Neurosurg Psychiat 49:1039–1043

Gennarelli TA, Thibault LE, Adams JH, Graham DI, Thompson CJ, Marcincin RP (1982) Diffuse axonal injury and traumatic coma in the primate. Ann Neurol 12:564–574

Graham DI, Adams JH, Doyle D (1978) Ischaemic brain damage in fatal non-missile head injuries. J Neurol Sci 39:213–234

Lindenberg R, Freytag E (1960) The mechanism of cerebral contusions. Arch Pathol Lab Med 69:440–469

McLellan DR, Adams JH, Graham DI, Kerr AE, Teasdale GM (1986) The structural basis of the vegetative state and prolonged coma after non-missile head injury. In: Papo I, Cohadon F, Massarotti M (eds) Le coma traumatique. Liviana Editrice, Padova, Italy

Viral Encephalitis and Its Pathology

J.R. Anderson

Introduction

The outcome of virus infections is determined by the interplay of viral pathogenicity, host susceptibility and host defence mechanisms. A wide variety of viruses are neurotropic in man, invading the central nervous system (CNS) through more than one portal and eliciting a spectrum of pathological changes, from florid acute necrotising inflammation to chronic degenerative disease. Certain neurotropic viruses, such as herpes simplex replicate, in all types of CNS cell. Others, such as the rabies virus, selectively attack neurons.

Host susceptibility is of fundamental importance. It is evident that a large proportion of people infected with neurotropic arboviruses never develop signs of neurological disease (GRIMLEY 1983). Similarly, only a tiny fraction of the population who are infected with herpes simplex virus develop encephalitis. Permissive viral infection in different CNS cell populations may be under host control, as exemplified by the ability of mouse hepatitis virus to induce encephalitis in certain strains of mice. Productive infection in neurons is governed by a single recessive mouse gene. The majority of neurons in resistant strains of mice fail to make infectious virus (KNOBBLER et al. 1981). The mechanism whereby these cells prevent viral replication is unknown, but it appears to operate after virus receptor interaction, as viral antigen is expressed in a very small proportion of the neurons in resistant mice (KNOBBLER et al. 1981). Selective cell vulnerability may be determined in similar ways in many neurotropic viral infections. Preliminary data comparing the distribution of viral genetic material in the CNS, as revealed by DNA hybridisation, with the distribution of viral antigens as shown by immunocytochemistry, suggests that non-productive infection is less cell specific and more widespread than permissive infection in a variety of different viral infections (BUDKA 1986a).

The viral properties of neuroinvasiveness and neurovirulence are equally significant. Neuroinvasiveness depends upon the ability of a virus to reach the CNS from a peripheral site, but the route taken varies. Thus rabies and herpes simplex virus are transported along axons (KRISTENSSON et al. 1982), whereas other viruses, such at HTLV-III (human T-lymphotrophic or leukaemia virus), appear to enter from the blood steam (BUDKA 1986a). Neurovirulence depends upon the ability of the virus to enter and usually to replicate within CNS cells. These properties may be encoded separately in the viral genome. Study of the Bunya virus family of arboviruses has revealed that neuroinvasiveness is the property of a segment of the RNA genome which normally codes for one non-structural and two envelope proteins. Bunya viruses with mutations in this segment show altered neuroinvasiveness (GONZALEZ-SCARANO 1986). Virus entry into a host cell is initiated by interaction of specific viral surface proteins and host cell membrane protein "receptors". Experimental investigation of reoviruses has identified segments of the viral genome responsible for this specific cell tropism. In the reoviruses the gene segment coding for a viral surface polypeptide which permits the virus to identify neurons, differs from that facilitating entry into ependymal cells (TYLER et al. 1985).

This chapter describes the pathology and pathogenesis of the most common forms of acute viral encephalitis and certain rarer subacute forms, where chronicity is related to host immunological deficit. Post-infectious encephalomyelitis will is also discussed, as it involves inflammatory destruction of CNS myelin triggered by a virus, although CNS viral invasion may not occur. Finally, slow virus or PRION infection has been included, although the exact nature of the infectious agent is not fully elucidated and its claims to be a virus have diminished since the original descriptions.

1 Acute Viral Encephalitis

1.1 Arboviral Encephalitides

1.1.1 Epidemiology

Collectively, arboviruses are responsible for the greatest number of cases of acute viral encephalitis throughout the world. This large group of some 450 enveloped RNA viruses is divided into 4 families according to antigenicity, but all are characterised by insect vectors (arthropod-borne), transmitted mostly by mosquitoes or ticks. These viruses vary in geographic distribution, but have considerable similarities in life cycles and patterns of human disease they inflict. Japanese B encephalitis virus (JBEV), Eastern equine virus (EEV) and St. Louis encephalitis virus (SLEV) result in the most severe encephalitides, with highest morbidity and mortality (LUBY 1979; GATUS and ROSE 1983), whereas others such as Western equine virus (WEV) and California encephalitis virus (CV) cause relatively mild, rarely fatal diseases (EMBIL et al. 1985) and are thus infrequently encountered by the histopathologist. JBEV is endemic in a vast geographical area in the far east, including not only Japan, but also China, Korea, Taiwan, Thailand and northern India and the territory appears to be expanding (GATUS and ROSE 1983). The equine viruses, SLEV and CV are the principal arboviral causes of encephalitis in the United States of America.

The arboviral life cycle involves viral replication in the primary insect host, transmission to and amplification within a second avian or mammalian, natural host. The virus is noncytopathic in both the insect vector and the natural animal resevoir. Man is an incidental second host, in whom the virus may cause disease, but viraemia is insufficient to continue the cycle. The arboviral encephalitides are characterised by seasonal incidence that is largely determined by appropriate climatic conditions and breeding grounds for the insect vector and frequently accompanied by high epizootic activity (LUBY 1979). Thus the JBEV is transmitted by culicine mosquitos which breed in rice fields and mosquito density is greatest in late summer. The virus appears to be neurotropic in the mosquito and viral antigen can be detected in retinal, thoracic and abdominal ganglion cells (JOHNSON et al. 1985). The neurons are persistently infected, but 1–2 weeks after invasion of the mosquito the virus reaches salivary glands and henceforth may be transmitted in saliva to animals or man. The natural reservoir includes small birds, but also the domestic pig (LUBY 1979). The migration of infected birds may widen the territory of arbovirus infection (GATUS and ROSE 1983), whereas a high infection rate in domestic pigs increases the chance of contact between man and carrier mosquitoes.

The maintenance of the arbovirus life cycle in the winter months in non-tropical climates is not fully explained, but may be due to persistent neural infection in hibernating female mosquitoes. With California virus transovarial transmission occurs, allowing overwintering in eggs and emergence of infected females in the spring. In regions where JBEV is endemic there is high population immunity. Children are the most vulnerable, but only about 1 in 200 of those who are bitten by infected mosquitoes will develop encephalitis. With St Louis virus, responsible for the greatest number of arboviral encephalitides in the USA, the elderly appear to be the most susceptible. After transmission of JBEV to a susceptible human host there is an incubation period of 6–16 days and a brief prodrome before clinical manifestations of encephalitis appear.

1.1.2 Entry of Virus into the CNS

The mode of entry of arboviruses into the central nervous system (CNS) is also uncertain. Experimental animal observations with SLEV indicate the olfactory pathway is the principal route. Intraperitoneal inoculation of Syrian hamsters resulted in only a transient mild viraemia, but a consistent rise in viral titre in the olfactory epithelium on the 4th day and sequential rise in titre in the olfactory bulbs and in the brain (MONATH et al. 1983). Haematogenous dissemination and invasion via CNS capillary endothelium has been proposed for other arboviruses (GRIMLEY 1983).

1.1.3 Pathological Changes

1.1.3.1 The Acute Disease

The pathology of arbovirus encephalitides varies in severity and CNS distribution, but the basic reactions of neural tissues are the same. The majority of pathological descriptions are obtained from autopsy studies in the most severe diseases, JBEV, EEV and SLV encephalitis. The death rate from JBEV encephalitis is estimated to be around 30%. When death occurs at the height of the disease the brain is grossly swollen and the leptomeninges may appear faintly cloudy. The cut surface shows dilated blood vessels and tiny greyish, foci of perivascular demyelination in the cerebral white matter may occasionally be visible to the naked eye. Histology reveals widespread inflammatory changes, principally in grey matter. There is perivascular lymphocytic cuffing, intense neuronophagia, foci of necrosis and clusters of inflammatory cells (GATUS and ROSE 1983; SHANKAR 1983). In white matter there may be foci of myelin breakdown. The leptomeninges contain a diffuse mononuclear inflammatory cell infiltrate. Inclusion bodies are not seen in arboviral encephalitides, although electron microscopy may reveal the viral particles (BASTIAN et al. 1975). Diagnosis in life depends upon early demonstration of IgM virus specific antibody in CSF or rising antibody titres in serum. At autopsy sites of maximal inflammatory changes in the brain may give some guide to the pathogen, but the patterns are not specific and there is considerable overlap. JBEV principally affects the basal ganglia, thalamus and brain stem. The clinical correlation is a Parkinsonian syndrome and early respiratory arrest. In the human brain JBEV antigen is only detected in neurons. Other arboviruses do not show this specificity. EEV has been demonstrated in oligodendroglia (BASTIAN et al. 1975). Investigation of the immune response to JBEV infection has shown an early T helper cell migration into the neural parenchyma accompanied by macrophages. By the 9th day after onset of encephalitis, viral antigen is almost entirely within macrophages, because neuronophagia has taken place. Antibody producing B cells appear within the CNS but stay confined within the perivascular spaces (JOHNSON et al. 1985).

1.1.3.2 Late Changes

Neurological sequelae in survivors of arboviral encephalitides are naturally greatest in the most severe forms and with JBEV are particularly seen in children. Permanent neurological deficit may occur and the pathological changes are still evident many years later. A small group of long term survivors all showed a similar pattern of lesions in the thalamus, substantia nigra and Ammon's horn (ISHII et al. 1977). The lesions were roughly symmetrical. Calcified and binucleated nerve cells were described in the thalamus. In addition there were small acellular, rarified areas, surrounded by gliosis and thought to correspond with the necrotic foci of the acute stage.

1.2 Herpes Simplex Encephalitis

1.2.1 Incidence

1.2.1.1 Children and Adults

HSV is the commonest cause of sporadic encephalitis in temperate parts of the world. The disease can strike at any age and affects males and females equally. Excluding neonatal infection, a large proportion of cases occur in childhood or adolescence; approximately 30% are under 20 years of age (NAHMIAS et al. 1983). HSV encephalitis of childhood and adult life is almost always attributable to HSV type 1 (NAHMIAS et al. 1982). However, concurrent with the increasing incidence of genital herpes, chiefly due to HSV type 2, there have been reports of HSV type 2 encephalitis in adults (Dix et al. 1985). Encephalitis is undoubtedly the most devastating manifestation of herpes simplex virus (HSV) infection in man. In untreated patients the mortality exceeds 70% and severe, permanent brain damage is common in survivors (Whitley 1984a). Fortunately, encephalitis is the least common pattern of herpes simplex infection, but when it does occur, it is characteristically an illness of sudden onset and rapid progression. In the USA it is estimated to occur in 1 per 750000 to 1 million individuals annually, approximately 250–500 cases in total (Whitley 1984a). In the UK, where absolute proof of infection from viral isolation is less often obtained, there are at least 50 cases, and possibly upto 200 cases per annum (Longson and Bailey 1980).

1.2.1.2 Neonates

Neonatal infection is almost invariably due to HSV type 2 acquired from the maternal genital tract at the time of delivery. Symptomatic genital tract lesions are unusual in the mother. In the USA Whitley has reported that at least 70% of the mothers of affected babies have no signs or symptoms of infection (Whitley and Alford 1982). Furthermore, the majority have no history of genital herpes or knowledge of contact with a symptomatic partner. Neonatal HSV infection inevitably has severe clinical manifestations. Surveys of normal infants at birth suggest that asymptomatic infection is exceptionally rare (Jenista 1983). Over one half of newborn infants with HSV infection are premature, possibly a reflection of increased susceptibility related to immaturity of the cellular immune system. At least 50% of infected neonates develop CNS infection. This may be localised encephalitis (15%) or associated with disseminated disease and multiple organ involvement, particularly liver, adrenals and lungs (31%) (Jenista 1983). There are an estimated 1000 cases of neonatal herpes per annum in the USA, where currently the disease appears to be more common than in the United Kingdom. Untreated, these cases have resulted in 50% fatality for localised CNS disease, at least 80% mortality for disseminated disease, and severe neurological deficit in many survivors (Whitley 1984a).

1.2.2 Primary or Secondary Infection

Beyond the neonatal period serological findings suggest that HSV encephalitis can be a manifestation of either a primary or recurrent infection (Nahmias et al. 1982). Primary infection is more common in the younger patients. Nevertheless, even in the older patients with known previous infection, encephalitis is not necessarily due to the same strain of HSV as that responsible for their recurrent mucocutaneous lesions. DNA restriction analysis of HSV isolates obtained from patients with concomitant encephalitis and mucocutaneous lesions and with serological evidence of previous infection, revealed genetically similar viruses in some cases, but two different strains in others (Whitley et al. 1982a). Unlike muco-cutaneous infection, triggering factors for HSV encephalitis have not been identified. Even immunosuppression, although associated with disseminated cutaneous lesions, does not appear to predispose to encephalitis (WHITLEY 1984b). HSV2 infection has occasionally been described in association with vasculitis and intracerebral haemorrhage in transplant recipients. HSV2 myelitis has been reported in a patient with AIDS (BRITTON et al. 1985) but it is not emerging as a frequent complication of this disease. Post-mortem reports of the acquired immune deficiency syndrome patients describe a variety of CNS infections, including cytomegalovirus and herpes zoster, but not herpes simplex encephalitis.

1.2.3 Clinical Features

1.2.3.1 Children and Adults

In childhood or adult infection HSV encephalitis is invariably an isolated organ infection, with characteristic predilection for the temporal lobes of the cerebral hemispheres. Concomitant herpes labialis or other skin lesions are occasionally present, but more often such lesions are absent. Initial non-specific symptoms of headache, fever and lethargy are usually proceed rapidly to behavioral disturbance, confusion and focal neurological deficit, such as dysphasia or hemiparesis (WHITLEY et al. 1982). Focal fits are a frequent manifestation in children, but less common in adults. Without treatment, the majority of patients follow a rapid downhill course. Deterioration in consciousness is accompanied by signs of raised intracranial pressure, progressing to coma, by which stage death is almost inevitable. Rarely, HSV encephalitis presents with a brain stem syndrome and cranial nerve palsies (DAYAN et al. 1972; ROMAN-CAMPOS and TORO 1980).

1.2.3.2 Neonates

Disseminated neonatal infection generally presents in the first week of life, whereas localised CNS disease may not be manifest until the second week (JENISTA 1983). Although skin vesicles develop in the majority (80–90%) of neonates, either at onset or during the course of their illness, in a few mucocutaneous lesions never appear (WHITLEY and ALFORD 1982). CNS disease is variously manifest as irritability, apnoea, focal or generalised convulsions, paralysis and eventual coma. Death may occur within hours to days, although some infants survive for several weeks.

1.2.4 Diagnostic Tests

In the early stages of HSV encephalitis neither clinical assessment nor routine tests can be relied upon for accurate diagnosis. Cerebrospinal fluid variables may be normal at first (KOSKINIEME et al. 1984) but there is usually a non-specific rise in the cell count after a few days. Specific antibodies are rarely detected in the CSF before the end of the first week of the illness and can still be absent towards the end of the second week. An increased ratio of CSF:serum HSV specific antibody is strong presumptive evidence of active CNS infection but can only provide late confirmation of diagnosis. Currently brain biopsy, to enable virus isolation from cerebral tissue, is the only means of early diagnostic proof (WHITLEY et al. 1982b; KOHL and JAMES 1985). With careful selection of biopsy site – and a temporal lobe is most often appropriate – there is a high yield, with very few false negatives (NAHMIAS et al. 1982). Following biopsy, even before culture results are available, a very rapid diagnosis, in a matter of hours, can be made in the majority by direct immunofluorescence of HSV antigen or by enzyme-linked immunosorbent assay (ELISA method; LAND et al. 1984).

1.2.5 Pathological Changes

1.2.5.1 Post-Mortem Findings

When death occurs within 1–3 weeks of onset of HSV encephalitis in children or adults, post-mortem characteristically reveals extensive, bilateral, assymetrical necrosis of the temporal lobes (Fig. 1). The lobes are soft, swollen and sometimes macroscopically haemorrhagic, due to red cell extravasation into the necrotic tissue (Fig. 2). Vascular injury and red cell leakage are late events in HSV type 1 encephalitis, there is no primary vasculitis. In fatal cases there is invariably cortical necrosis in other regions of the cerebral hemispheres, notably the frontal lobes and cingulate gyri, but the damage is usually maximal in the temporal lobes. The brain stem is largely spared, apart from the very rare cases with a clinical brain stem syndrome, in which this region is severely

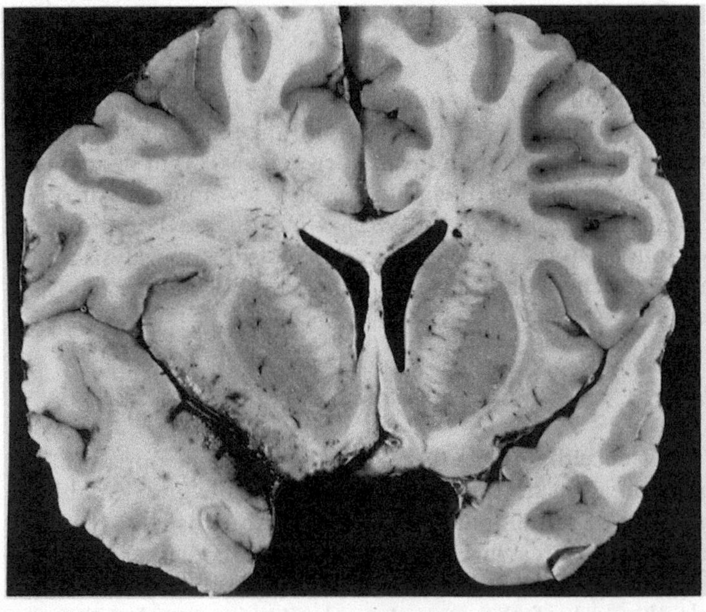

Fig. 1. Acute herpes simplex encephalitis in a 27 year old man who became drowsy during the course of a mild pyrexial illness. He lapsed rapidly into coma and death occurred 3 days later. Asymmetrical swelling and haemorrhagic necrosis of the left temporal lobe

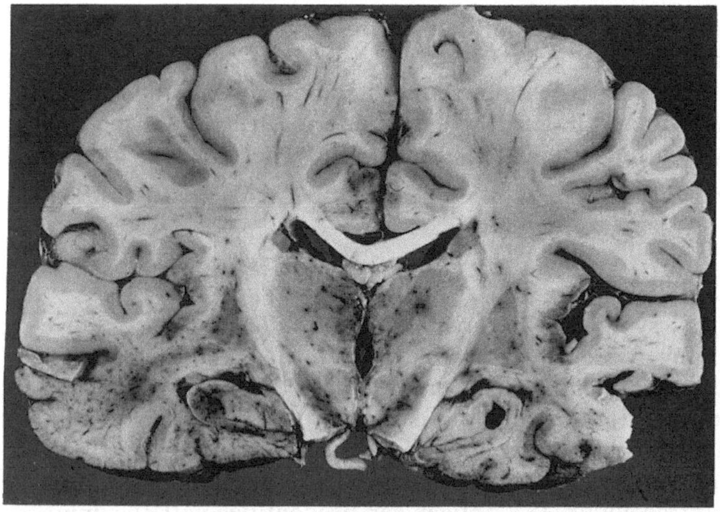

Fig. 2. Acute herpes simplex encephalitis in an elderly female. Bilateral narrowing of the cortex of the parahippocampal gyri due to acute necrosis and blotchy appearance of the temporal lobes due to dilated blood vessels and tiny haemorrhages

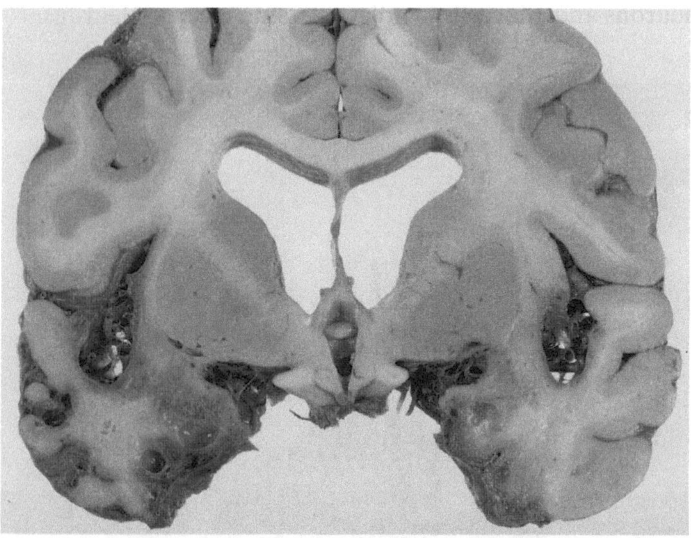

Fig. 3. Long term survivor of herpes simplex encephalitis. Destruction of the cortex and cystic change in the subcortical white matter of the left temporal lobe. Changes are similar, but less severe on the right. The cortex of the insula is also severely narrowed

and sometimes exclusively affected (Dayan et al. 1972; Roman-Campos and Toro 1980). Occasionally the virus may spread from the CNS along the optic nerves to the eye, eliciting a necrotising retinitis (Minkler et al. 1976; Johnson and Wisotzkey 1977). In the minority of patients who survive and come to post-mortem months or years later the temporal lobes are severely shrunken with gross thinning and yellow discolouration of the cortex and cystic changes in the underlying white matter (Fig. 3).

In neonatal HSV encephalitis temporal lobe predominance is unusual and cerebral necrosis is widespread. Extensive white matter destruction in the cerebral hemispheres may result in encephalomalacia and parenchymal calcification (BENATOR et al. 1985).

1.2.5.2 Initial Cellular Events and Correlation with Clinical Progression

The evolution of clinical symptoms in most childhood and adult cases corresponds with an enlarging focus of virus replication in a temporal lobe. The ensuing necrosis, inflammation and oedema are responsible for diminishing consciousness and signs of raised intracranial pressure.

The cellular events have been closely studied in human biopsy and postmortem material (ESIRI 1982) and in animal models of HSV encephalitis (ANDERSON and FIELD 1983). Intracellular HSV antigen can be identified with the immunoperoxidase technique in both nucleus and cytoplasm of infected neurons and glial cells at least several hours before there is any cellular infiltration (Plate 1A, see p. 39). The initial focus of active virus infection, usually in the temporal lobe, rapidly expands. HSV spreads from cell to cell, creating large, confluent foci of infected cortical neurons and glial cells in adjacent white matter. Neurons,

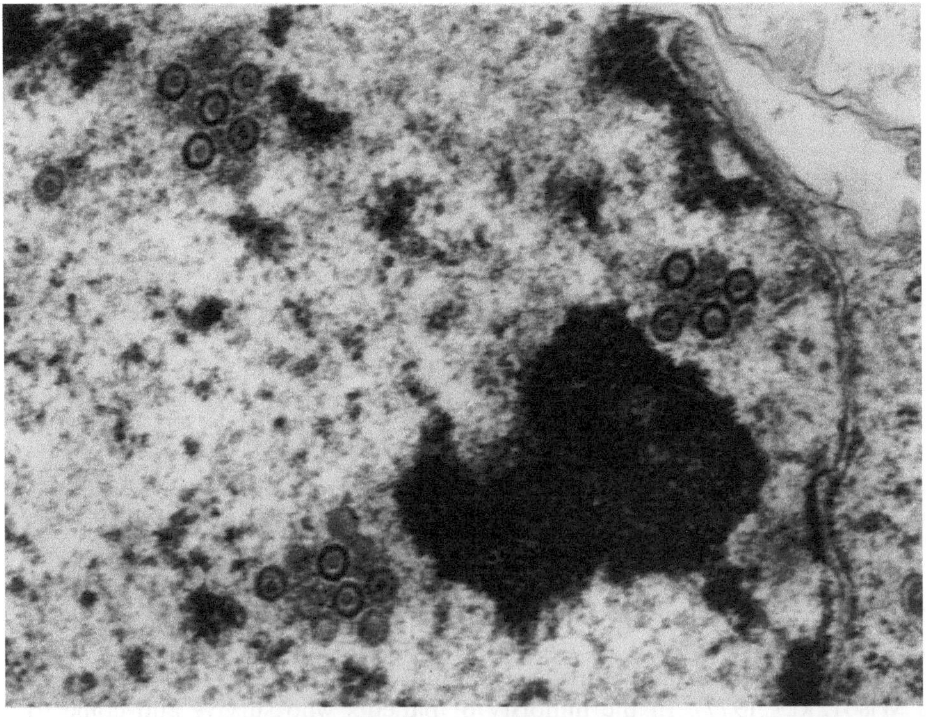

Fig. 4. Herpes simplex encephalitis. Intranuclear nucleocapsids revealed by electron microscopy

astrocytes and oligodendrocytes all show cytolytic infection with HSV, but the dentate fascia of the hippocampus appears to be particularly susceptible (Plate 1B). The apparent vulnerability of this layer may be due to a specific property of these large neurons, such as a high number of surface receptors, but intense infection may only be a reflection of the very closely packed cell bodies which readily permit cell to cell spread. The earliest inflammatory response is a transient influx of polymorphoneutrophil leucocytes at a time when neurons are beginning to appear abnormally eosinophilic and slightly shrunken. Eosinophilic intranuclear inclusions may be found (Plate 1C), but they are far less conspicuous than the brightly stained inclusions of SSPE. Nuclei of infected cells without stainable inclusions may show margination of the chromatin and an 'empty' nucleus. Nucleocapsids can be demonstrated in both nuclei and cytoplasm by electron microscopy (Fig. 4). The initial neutrophilic response is rapidly followed by the more conspicuous, perivascullar infiltrate of lymphocytes and plasma cells, which coincide with obvious cell death, disintegration and neuronophagia. By the time necrosis and inflammation reach a peak, often about 7–10 days after onset in man, detectable virus antigen is waning. At this stage the widespread cell death in large areas of temporal cortex, combined with breakdown of subcortical white matter and massive influx of macrophages resembles a recent infarct (Fig. 5).

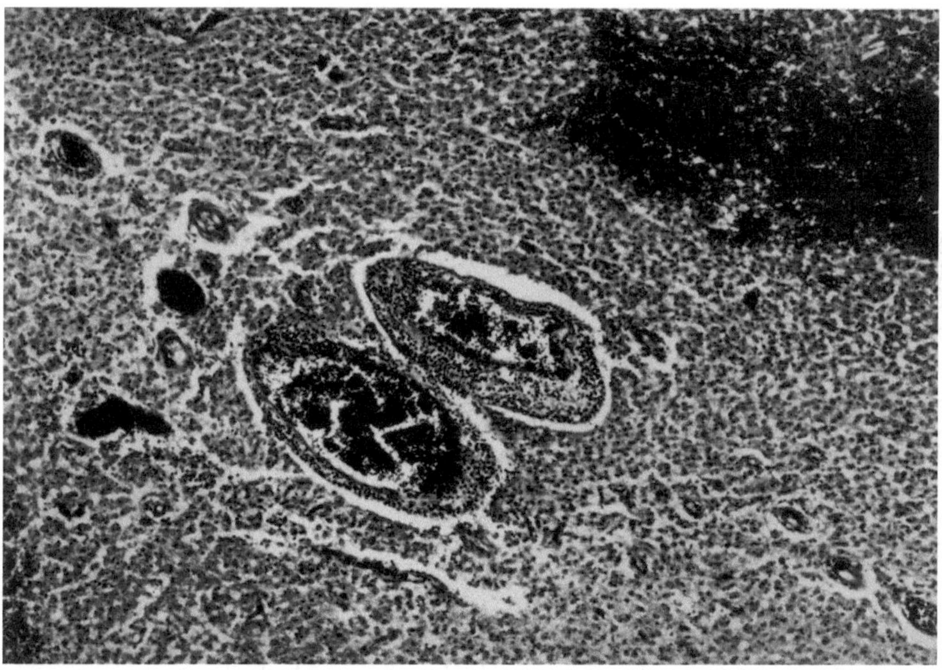

Fig. 5. Necrotic haemorrhagic temporal cortex in acute herpes simplex encephalitis. There is total destruction of neurons, massive infiltration by macrophages and perivascular cuffing by mononuclear chronic inflammatory cells. A small perivascular haemorrhage is typical of the advanced phase of virally induced necrosis and inflammation

1.2.6 *Entry of Virus into the CNS*

Post-mortem studies of HSV encephalitis in man have emphasized the temporal lobe location, with involvement of cell groups which constitute the limbic system. With the knowledge that HSV spreads from cell to cell and can travel along axons, this distribution strongly suggests that the olfactory nasal mucosa is the portal of entry of virus into the central nervous system. The primary olfactory neurons originate in the nasal mucosa, synapse in the olfactory bulbs on the inferior surface of the frontal lobes and connect with the limbic structures of the temporal lobes. The hypothesis is supported by postmortem studies in man, which have shown HSV antigen in the olfactory pathway and limbic system, but little elsewhere (TWOMEY et al. 1979; ESIRI 1982; OJEDA et al. 1983) and by a distribution of infection following intranasal inoculation of HSV in mice (ANDERSON and FIELD 1983). The vulnerability of the olfactory pathway may be explained by its anatomical peculiarity. The primary olfactory neuron is unique in that it is sheathed by astrocytic cells (BARBER and LINDSAY 1982). The peripheral portions of all other cranial nerves (except the optic nerve) are encased by Schwann cells. Astrocytes are susceptible to HSV infection, whereas infection is usually abortive in Schwann cells. The primary olfactory neuron can thus be likened to an extension of the central nervous system. Although the presence of latent HSV in the trigeminal ganglion is well documented in man, it seems that in the majority of cases, encephalitis is not due to reactivation and spread along the central portion of the fifth nerve. This is the likely route only in the very rare cases of brain stem encephalitis.

Recognition of the portal of entry does not, however, explain why this ubiquitous virus actively infects the central nervous system in but a tiny fraction of the population exposed to it. There is no evidence that the viruses isolated from cases of encephalitis have had unusual neurovirulent properties and epidemics of HSV encephalitis do not occur. Therefore it is likely that host factors determine susceptibility. Immune responsiveness may have some role, but the infrequent appearance of HSV encephalitis as opposed to cutaneous lesions in the immunosuppressed implies some other tissue specific factor. There is some speculation that HSV can establish latency in intracerebral neurons as well as in trigeminal ganglion cells. Stable HSV specific DNA sequences have been detected in the CNS of mice after recovery from acute infection mice and have been shown to have the same organisation as sequences in human trigeminal ganglia (EFSTATHIOU et al. 1986). The evidence for the presence of HSV DNA in the human brain is slender, but small amounts would be difficult to detect. However, even if latent infection of human intracerebral neurons does occur, the sudden conversion from latent to permissive infection is unexplained.

In neonatal encephalitis the widespread necrosis in the brain suggests a direct blood borne infection. A case of necrotising vasculitis has been described in disseminated neonatal infection with HSV type 2, in which electron microscopy confirmed viral replication in arterial endothelial cells (PHINNEY et al. 1982). Animal experiments also offer supportive evidence. In suckling mice, after intraperitoneal inoculation, HSV antigen was initially distributed in and around small blood vessels, whereas in older animals, both intravenous and intranasal inoculation gave a consistent pattern of involvement of certain neuron groups, but no vascular lesions (JOHNSON 1964).

1.2.7 Prognosis

The prognosis of HSV encephalitis has dramatically improved since the introduction of Acyclovir (ACV; SKOLDENBERG et al. 1984). This drug can eliminate the virus from CNS tissue (FIELD et al. 1984). However success of this antiviral chemotherapy measured in terms of quality of life in survivors depends not only upon the proven efficacy of this drug, but upon institution of therapy at an early stage before widespread neuronal destruction has occurred. A small percentage of patients with HSV encephalitis survived before the advent of specific antiviral therapy and invariably suffered from neurologic or neuropsychologic sequelae (RENNICK et al. 1973), correlating with the temporal lobe destruction described above. An amnestic syndrome and bizarre behavioral disturbances are recorded in brain damaged survivors of HSV encephalitis. The efficacy of ACV is such that many more semicomatose and even comatose patients may now survive. Complete recovery from advanced disease is impossible and residual neurological deficit inevitable. In the future physicians may encounter these late neuropsychological complications more frequently (GREENWOOD et al. 1983) and pathologists may see more cases of old, patchy temporal lobe destruction at post-mortem.

2 Post-Infectious – Perivenous Encephalomyelitis

Acute, monophasic neurological disease following rapidly upon a common viral illness is a well recognised clinical syndrome with the pathological basis of CNS perivenular demyelination. Historically, similar pathology was first recorded as a complication of rabies vaccination. The discovery that this was due to contamination of the vaccine by animal CNS tissue, rather than transmission of a neurovirulent virus, paved the way for the many animal models of experimental allergic encephalomyelitis (EAE; RIVERS and SWENTKER 1935). Perivenular demyelination was also shown to be the basis of serious neurological sequelae of small pox vaccination. Subsequently, this same pathological process has been linked with neurological complications of a wide variety of viral infections, including measles, mumps and chicken pox. Although the interval between acute viraemia and appearance and severity of the neurological symptoms is variable, the pathology provides unity. Neurological disease is triggered by a viral infection, but the perivenular demyelination that occurs cannot be attributed to the direct cytopathic effects of the virus. In many instances virus cannot be detected in the CNS and as in post-rabies vaccination (TORO et al. 1977; GAMBOA et al. 1983) and EAE, myelin breakdown is considered to be due to an allergic phenomenom (JOHNSON et al. 1984). Post-infectious, parainfectious or perivenous encephalomyelitis (PVE) and acute disseminated encephalomyelitis are among the many clinical and pathological terms that all refer to this same condition. Acute haemorrhagic leucoencephalitis describes the pathology of a fulminating, invariably fatal CNS disorder that often follows an unspecified, but presumed viral infection of the upper respiratory tract. Although the pathology is more acute there is overlap with PVE and it is probably a florid example of postinfectious allergic CNS injury (Johnson 1984; Boos and Esiri 1986).

With the eradication of small pox, complications of vaccination have disappeared and measles infection is now the commonest cause of PVE. Although the incidence of measles infection has been greatly reduced by mass vaccination in Europe and North America, there are still many parts of the world where it is endemic and periodic epidemics occur. World-wide, 1.5 million children of measles every year. PVE is estimated to occur in 1 in 1000 cases of measles infection (Johnson 1984). It is unpredictable. There is no link between the severity of the initial exanthem and the development of PVE. No genetic predisposition or HLA association has been discovered. The latent period between onset of the rash and neurological symptoms is usually about 5 days, but can be longer. PVE induced by measles has a high mortality, up to 40% and there is a high incidence of neurological deficit in survivors.

2.1 Macroscopic Appearances and Histology

2.1.1 Perivenous Encephalomyelitis

Death due to PVE is usually within a week or two of onset and the pathological changes are similar irrespective of the primary viral infection. Macroscopic changes are not distinctive, but the brain is likely to show some degree of symmetrical swelling and congestion of the leptomeninges. Dilated small blood vessels are often conspicuous in the cut surface of the white matter and a few petechial haemorrhages may be visible. The ventricles appear small due to the oedema. Histology shows small venules throughout the white matter cuffed chronic inflammatory cells and surrounded by a narrow sheath of demyelination (Fig. 6; Toro et al. 1977). Inflammatory cells in the perivascular cuff are predominantly macrophages and lymphocytes, the latter mainly cytotoxic suppressor T cells (Booss and Esiri 1986). Plasma cells are scanty and it is of note, that in contrast with SSPE, in post-infectious measles encephalitis no anti-viral antibody is detected in the CSF. The number of macrophages and degree of cytoplasmic lipid accumulation depends upon the duration of the illness. Silver stains show there is relative sparing of axons, but some tortuosity and fragmentation will be found in the demyelinated zone. The central venule may show plump endothelial lining cells, scanty red cell extravasation and polymorphoneutrophil emigration, but not the nuclear dust and fibrin extravasation of an acute vasculitis. In long term survivors, whose brains have been examined many years later, the inflammation has disappeared, but as would be expected in the CNS evidence of demyelination persists and is associated with dense perivascular gliosis (Booss and Esiri 1986).

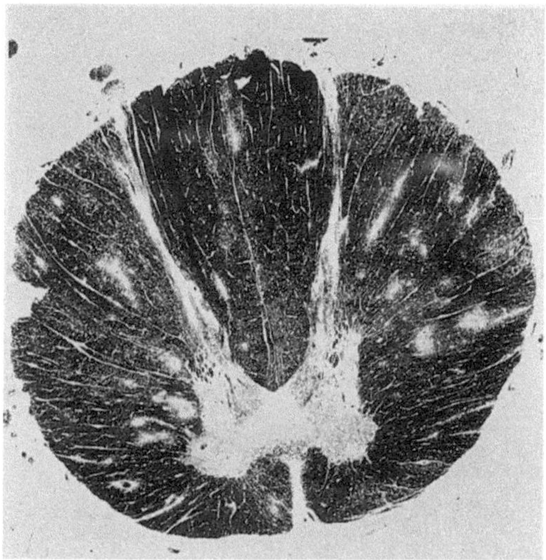

Fig. 6. Post-infectious encephalitis. Perivascular demyelination in the spinal cord of a patient dying shortly after an attack of influenza. (Weigert-Pal stain for myelin)

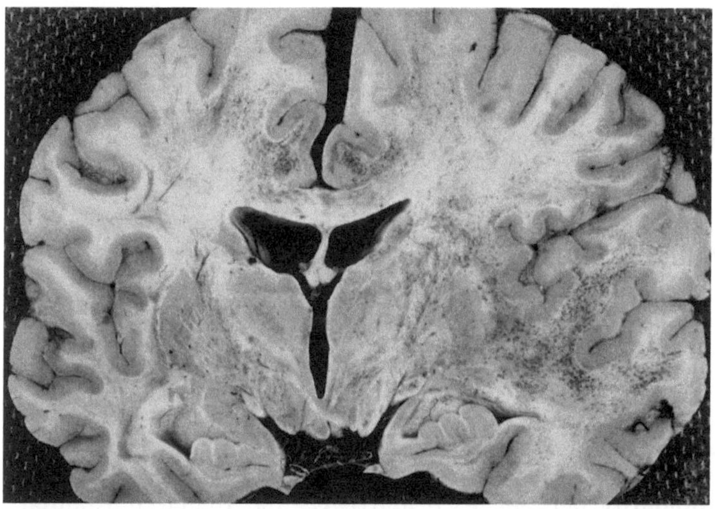

Fig. 7. Acute haemorrhagic leucoencephalitis in a 30 year old, who became comatose and died within in few days of onset of a mild upper respiratory infection. There are numerous tiny petechial haemorrhages dispersed through the white matter, but the cortex is unremarkable

2.1.2 Acute Haemorrhagic Leucoencephalitis

In acute haemorrhagic leucoencephalitis the macroscopic changes are similar, but tiny petechial haemorrhages are far more numerous and scattered diffusely within the white matter where they impart a pinkish discolouration to the cut surface (Fig. 7). Histology shows dilated capillary – venules with obvious red cell extravasation and small perivascular collections of inflammatory cells

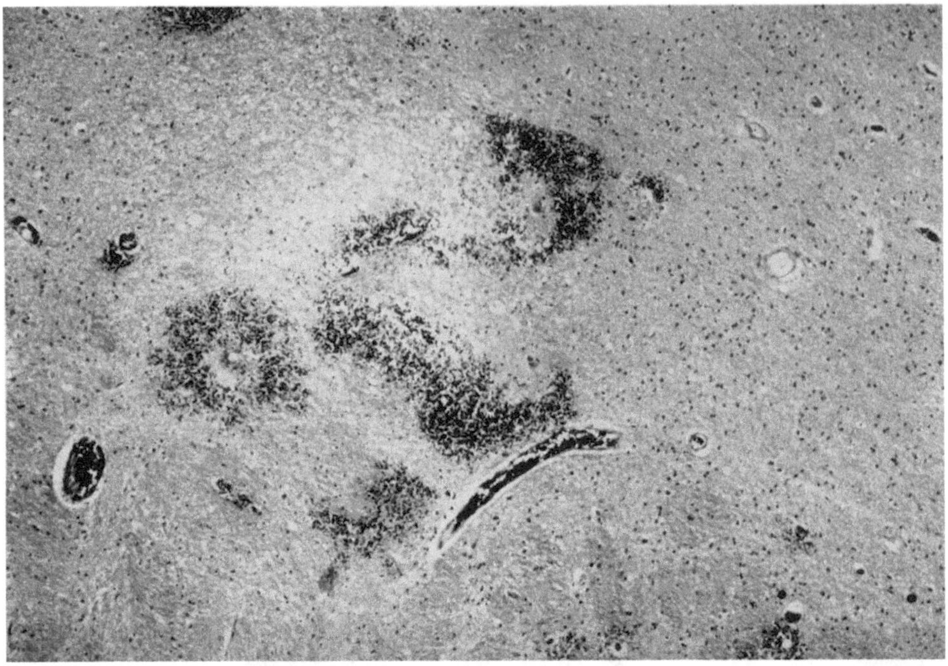

Fig. 8. Acute haemorrhagic leucoencephalitis. Small vessels in the white matter are surrounded by extravasated red cells and acute inflammatory cells

(Fig. 8), but in contrast to PVE, polymorphoneutrophil leucocytes are in excess of mononuclear cells (Plate 1D). A narrow zone of surrounding white matter is pale and oedematous, with early signs of myelin fragmentation, but in this catastrophic illness, death supervenes before there is significant demyelination. Foamy macrophages therefore do not appear.

2.2 Pathogenesis

Despite histological similarity between PVE and EAE and the circumstantial evidence to support an allergic disease in man, the pathogenesis is not entirely

Plate 1A. Cortical neurons and some satellite oligodendroglial cells containing herpes simplex virus ▷ antigen. Virus antigen is present in nuclei and cytoplasm and also in proximal axons and nerve cell processes. (Immunoperoxidase method to demonstrate herpes simplex virus antigen)

Plate 1B. Confluent antigen containing cells in the dentate fascia of the hippocampus in a case of acute herpes simplex encephalitis. Numerous oligodendrocyte nuclei in the white matter are also stained. (Immunoperoxidase method to demonstrate herpes simplex virus antigen)

Plate 1C. Herpes simplex encephalitis in a 3 year old girl. Section of cerebral cortex showing eosinophilic intranuclear inclusion bodies, prior to the appearance of the inflammatory response

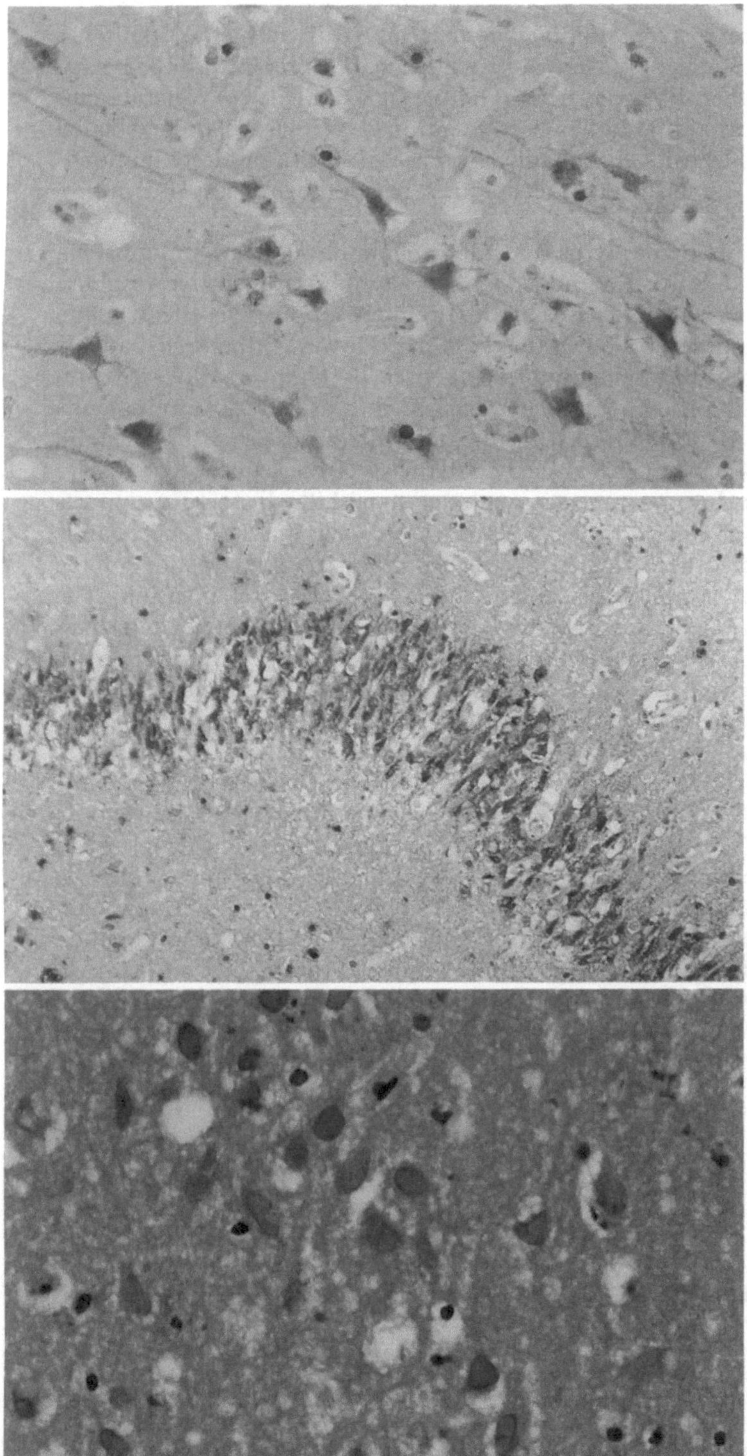

A B C

clear. Although there is a possibility that measles virus may enter the CNS in the course of acute infection there is no evidence of its replication or persistence in CNS cells when PVE develops (GENDELMAN et al. 1984). The absence of intrathecal humoral antibody has been mentioned. No virus can be isolated from the brain in postmeasles encephalitis, or detected in situ by either immunohistochemical or DNA hybridisation techniques. The pathogenesis is thus entirely different from that of SSPE.

Measles is a lymphotropic virus and all patients develop a lymphopaenia in acute infection and show a depression of the lymphoproliferative response to mitogens. There is no evidence that this immunosuppressive effect is any greater in those who develop PVE (HIRSCH 1984). However, the PVE patients do appear to develop aberrant immune responses. A lymphoproliferative response to myelin basic protein can be demonstrated in at least 50% of patients with PVE, but only 10% of those with uncomplicated measles virus infection (JOHNSON et al. 1984). It can be argued that this is a consequence of myelin breakdown, but its early appearance, in conjunction with the histology, supports a role for cytotoxic T lymphocytes. Thus measles virus induced encephalitis is an autoimmune disorder, most probably resulting from interaction of the virus with peripheral lymphocytes and not due to CNS viral invasion. The same mechanism may operate with other viruses that initially replicate in lymphocytes, but why there is selective attack upon a CNA antigen is unexplained. If completely different viruses can act in this way, molecular mimicry between viral and CNS antigens seems unlikely, but shared CNS and lymphocytes antigens could be relevant.

The mechanisms of virus induced demyelination are not necessarily the same in every example of PVE. Other viruses believed to initiate PVE have been identified in the CNS, at least in small amounts. Mumps is a neurotropic virus and has been cultured from CSF whilst the patient has acute parotitis (JOHNSON 1984). In post-vaccinal encephalomyelitis viral invasion of the CNS is thought to precede demyelination. Experimental animal models of CNS demyelination illustrate the diversity of mechanisms of virus induced demyelination (MARTIN and NATHANSON 1979). Thus viruses, other than measles, may act through antigenic modulation of CNS antigens or through possession of common antigenic determinants with CNS tissue.

Plate 1D. Acute haemorrhagic leucoencephalitis. Small venule surrounded by polymorphoneutrophil ▷ leucocytes and extravasated red cells. (Haematoxylin and eosin)

Plate 1E. Eosinophilic intranuclear inclusion bodies in SSPE. (Haematoxylin and eosin)

Plate 1F. Subacute sclerosing panencephalitis. Intranuclear inclusions are present in oligodendroglia and neurons, but are not seen in the surrounding reactive astrocytes. (Immunoperoxidase method to demonstrate glial fibrillary acidic protein)

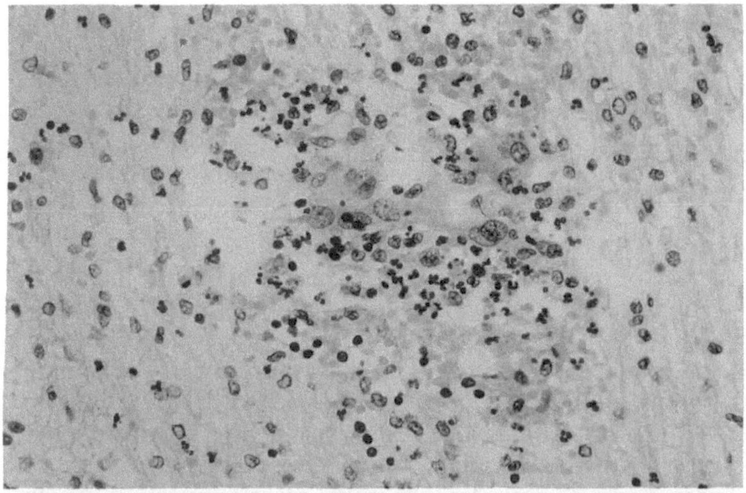

D

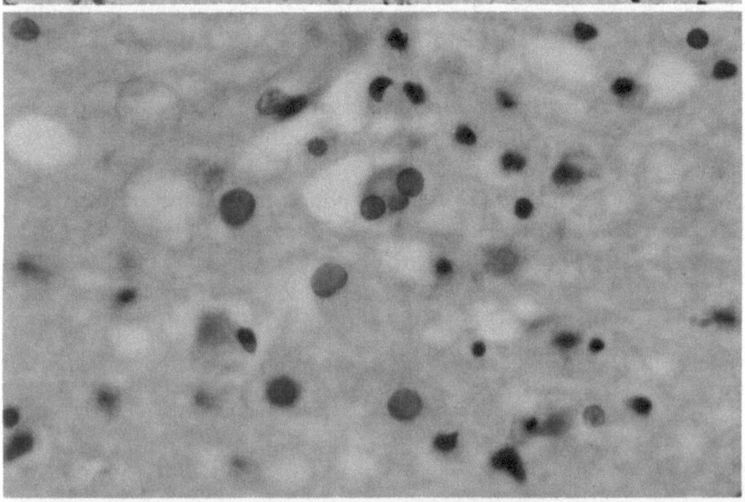

E

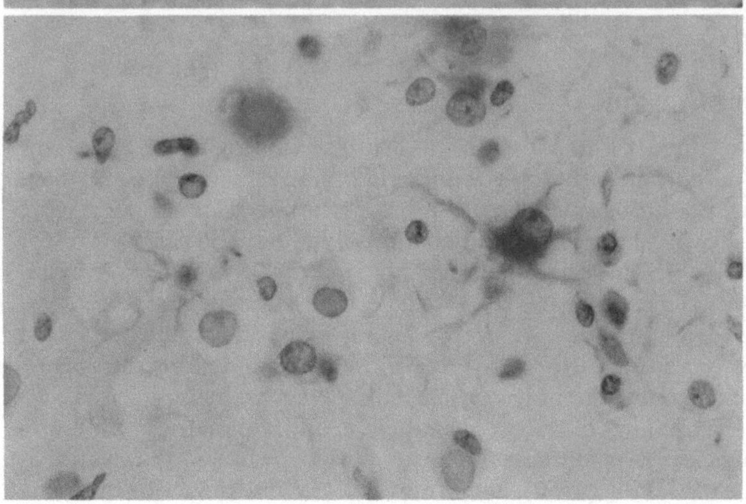

F

3 Persistent Infection

3.1 Subacute Sclerosing Panencephalitis

SSPE is a rare, inevitably fatal late complication of measles virus infection that results from persistence of the virus in the CNS. In the USA it is estimated that SSPE will develop in between 6–22 of every million children who have an acute measles infection (AABY et al. 1984). The advent of a measles vaccination programme in the USA brought about an immediate decline in the incidence of acute measles infection and after a lag of several years the number of cases of SSPE also fell (GRAVES 1984). The controversy of whether SSPE is ever induced by the vaccine is not resolved, but it is clear that any risk is minute in comparison with that of natural infection (GRAVES 1984). Epidemiological studies disclose that SSPE is 2–4 times more common in boys than girls and victims usually live in rural areas (AABY et al. 1984). The clinical manifestations of SSPE appear between 4 and 7 years after the initial measles infection, which has most often occurred at an exceptionally young age (SEVER 1983). Thus the average age of onset of SSPE is 7–8 years. The onset of neurological disease is insidious, but progressive deterioration of all higher neurological function leads inevitably to coma and eventually death within 1–3 years of onset. A minority of patients progress very rapidly and die within 6–8 weeks of presentation (GILDEN et al. 1975; SILVA et al. 1981), whereas an equally small percentage have a chronic, fluctuating course, deteriorating slowly over 5–10 years. Longer survival and even periods of spontaneous improvement have occasionally been reported (RISK et al. 1978), but it is doubtful if complete remission has ever occurred. There is a high serum titre of measles specific antibody, greater than in acute measles infection (SEVER 1983). Oligoclonal IgG measles antibody is present in CSF and CSF/serum antibody ratios indicate intrathecal antibody synthesis, a phenomenom that does not occur in the usual acute infection (JOHNSON 1984).

3.1.1 Pathological Changes

3.1.1.1 Macroscopic Appearances

Irrespective of the tempo of the disease, at death, pathological changes are widespread in the cerebrum, in both grey and white matter. However, the macroscopic abnormalities at post-mortem are more pronounced in long survivors (BOOSS and ESIRI 1986). A greyish leptomeningeal exudate is generally present at all stages and may be the only external abnormality in fulminating cases. In typical cases, with clinical duration of at least a year, some degree of cerebral atrophy is usually evident. There is narrowing of cerebral gyri and at least patchy thinning of the cortical ribbon. In the most chronic cases atrophy may be severe with conspicuous cerebral cortical narrowing, reduction in volume of deep white matter, thinning of the corpus callosum and symmetrical dilatation of the lateral ventricles. In such cases, as exemplified by the 7 year survivor reported by MALAMUD et al. (1950), there may be a gross reduction in brain weight. In the cut surface loss of myelin is visible as ill defined grey areas in cerebral white matter, patchily distributed in cases of short duration, progressing to diffuse bilateral change in the deep white matter of long survivors. Both cerebral cortex and white matter develop an increasingly abnormal firm texture due to gliosis (BOOSS and ESIRI 1986).

3.1.1.2 Histopathology

At any stage histology is practically pathognomonic. The key feature is the presence of intranuclear inclusion bodies in neurons and oligodendroglia. With ordinary H&E staining these inclusions appear as homogeneous, round eosinophilic structures displacing nuclear chromatin into a barely visible, peripheral basophilic rim (Plate 1E). A clear halo may separate the inclusion and the chromatin. Inclusion bodies are numerous and even abundant in both rapidly progressive cases and typical cases of intermediate survival. Their distribution correlates with areas of maximal tissue damage, being greater in cerebrum than brainstem and cerebellum and least in spinal cord. Inclusions are still present, although relatively scanty in the most chronic cases, and may only be seen in neurons. Immunohistochemical techniques confirm that inclusion bodies contain measles virus specific antigen. Positive staining is also seen in nuclei and cytoplasm of many neurons and oligodendrocytes that do not contain obvious inclusion bodies, but reactive astrocytes are entirely negative (ESIRI et al. 1981; Plate 1F). Residual neurons may appear degenerate and shrunken, often undergoing neuronophagia. Acute neuronal death and karyorrhexis may be conspicuous in fulminating cases (SILVA et al. 1981; Fig. 9), whereas complete neuronal dropout is obvious and sometimes associated with cortical spongiosis in the chronic cases (MALAMUD et al. 1950).

Accompanying the inclusions are the inflammatory changes common to most classical CNS viral infections, qualitatively similar, but quantitatively less than acute viral encephalitides. There is a scattered leptomeningeal infiltrate of lymphocytes, histiocytes and plasma cells. Perivascular cuffing by these same cell types is widespread in cerebral cortex, white matter and basal ganglia and may also be found in the brainstem, cerebellum and spinal cord, particularly in the more chronic cases. The degree of inflammation is variable, but is usually greatest in cases of shortest duration. Plasma cells are the source of intrathecal antibody and immunoperoxidase staining confirms that the majority are producing IgG (ESIRI et al. 1981). In white matter, secondary to oligodendrocyte infection, there is myelin loss, initially patchy, later extensive and accompanied by some degree of axonal degeneration and by astrocytic gliosis. Demyelination is maximal in the deep central white matter and corpus callosum. The pyramidal and optic pathways are relatively spared. In the early stages of myelin breakdown foamy, lipid filled macrophages are numerous, but later retreat to the perivascular spaces. Reactive astrocytic gliosis develops alongside the degenerative changes, in both cortex and white matter, becoming intense in the white matter of long survivors. Comparison of the distribution of CNS damage in long and short term survivors suggested the virus may spread slowly through the CNS in a caudal direction.

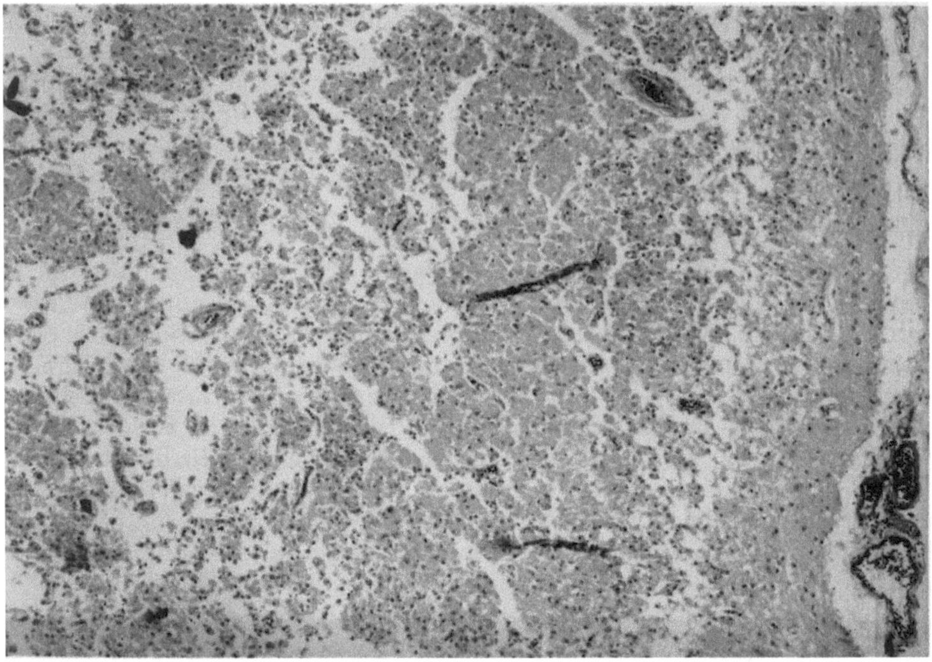

Fig. 9. Subacute sclerosing panencephalitis. A rapidly progressive case of only 8 weeks duration in a nine year old boy who had measles at age 3 years. Severe loss of neurons, with many cells in the process of degeneration in the cerebral cortex. Inclusion bodies were numerous

3.1.1.3 Electron Microscopy

Electron microscopy reveals "smooth" myxovirus nucleocapsids, predominantly located within nuclei (BOOSS and ESIRI 1986). In comparison in acute, wild type measles many nucleocapsids, described as "fuzzy" are seen in the cytoplasm and budding from the surface of infected cells. The fuzziness is due to attachment of matrix – M protein, an essential prerequisite for budding. In SSPE a fundamental abnormality is the complete absence of budding from CNS cells and mature virions are never observed. Although fuzzy cytoplasmic nucleocapsids are sometimes identified in fulminant SSPE (SILVA et al. 1981) they do not show the alignment beneath the cell membrane that normally precedes budding (JOHNSON 1984).

3.1.1.4 Neurofibrillary Tangles and Dendritic Abnormalities

In a few of the most chronic cases of SSPE neurofibrillary tangles, morphologically identical to those in Alzheimer's senile dementia, have been observed (MALAMUD et al. 1950; PAULA-BARBOSA et al. 1979). These were not accompanied by senile plaques. Golgi preparations on cortical biopsies taken early in the clinical course of SSPE have shown a paucity of dendritic spines on pyramidal neurons (PAULA-BARBOSA et al. 1980). Electron microscopic studies suggest that mitochondrial abnormalities may precede degeneration of the dendritic spines.

Dendritic abnormalities are also described in Alzheimer's disease and in mentally retarded patients. Thus, in SSPE it is postulated that changes in cortical dendrites are the morphological correlate of the characteristic insidious early dementia (Paula-Barbosa 1984).

3.1.2 Pathogenesis

The pathogenesis of SSPE has not been fully elucidated, but many of the threads of information can be woven into a hypothesis, to explain why the virus persists and why it causes a chronic neurologic disease.

3.1.2.1 Hypothesis

Initial persistence of measles virus in the CNS is due to the combined effects of immunological immaturity of the host, persistence of maternal antibody, relative non-permissiveness of CNS cells and a high infective dose. Once viral persistence is established, wild type measles virus is converted to a defective cell associated virus, which is not recognised by the host immune system. Free virions are not produced and the virus spreads through the CNS by cell to cell fusion. Gradual intracellular build up of virus material eventually has a cytopathic effect culminating in cell death.

3.1.2.2 Evidence

Epidemiological studies in man indicate that the primary measles virus infection occurs at an exceptionally young age, often under 2 years and suggest that heavy exposure is important (Aaby et al. 1984). The early age of primary infection implicates immunological immaturity and the presence of residual maternal antibody. Experimental animal studies reinforce the importance of these factors. In hamsters measles virus inoculation causes a fatal encephalitis in neonates, but persistent infection of neonates born to immunised mothers (Johnson 1984). Maternal antibody may impede virus elimination by blocking the afferent arm of the immune response. Effective cell mediated immunity may be diminished through measles virus infection of T lymphocytes, but there is no evidence of any primary immunological deficit in the children who develop SSPE (Johnson 1984). Transient neurological disturbance with EEG changes occasionally complicate the common childhood exanthem and it is possible that viraemia carries the virus to the CNS during the primary infection (Graves 1984). However, a recent study employing in situ hybridisation and a cloned measles virus DNA probe has detected measles virus RNA in a high percentage of peripheral blood lymphocytes from SSPE patients (Fournier et al. 1985). An alternative theory is that persistent infection is initially established in lymphocytes, a major target of the primary infection and at a later time, perhaps because of some alteration in the blood brain barrier lymphocytes carry the virus into the brain. Certain CNS cell lines are non-permissive in vitro, thus it is quite possible that human CNS cells likewise restrict viral replication in vivo (Choppin 1981).

In SSPE although measles virus antigen is identified by immunocytochemistry and paramyxovirus nucleocapsids are seen by electron microscopy, the virus cannot be recovered by conventional culture, only by co-cultivation with permissive cells. There are reduced amounts of the viral absorption and fusion proteins normally inserted into the host cell membrane. M protein is essential to viral replication and appears in all permissive cells. It is inserted into the cell membrane where it stabilises the cell surface viral proteins and provides the attachment site for cytoplasmic nucleocapsids prior to budding (JOHNSON 1984). In infected CNS cells either the virus M protein is not synthesised or an abnormal M protein is produced (SHEPPARD et al. 1985).

The absence of any clusters of SSPE strongly indicate that the initial infection is not due to a neurovirulent strain, but to classical measles virus which becomes modified within the host. The emergence of a mutant virus, which cannot direct synthesis of normal M protein, is encouraged by persistent infection, because the paramyxoviruses have a naturally high mutation rate (SHEPPARD et al. 1985). Failure of the mutant to express antigens on the cell surface hides the virus and gives protection against the host cytotoxic T cells. In SSPE where the normal budding mechanism fails, the virus can only spread by cell to cell fusion and this is probably a slow process because of the paucity of fusion protein. Accumulation of intranuclear nucleocapsids eventually leads to cell death and release of virus protein which chronically stimulates host immunity and explains the high antibody titres. Experimental studies suggest that intrathecal antibody may further impede cellular immune attack and promote persistence (JOHNSON 1984). In vitro exposure of measles virus infected cells to excess antibody can modulate viral surface antigens. Essentially antibody combines with and removes viral surface antigens, thus preventing T cell recognition. In the absence of complement fixation the antibody itself has no cytolytic effect.

4 Opportunistic Infections

4.1 Acquired Immune Deficiency Syndrome – AIDS

The aetiologic agent of the acquired immune deficiency syndrome is a retrovirus, HTLV-III (HIV Human immunodeficiency virus). Viral destruction of T helper cells produces a characteristically severe decrease in the normal T helper-T suppressor cell ratio, with resultant multiple opportunist infections, including CMV, pneumocystis, mycobacterial and fungal infection and neoplasia in the form of B-cell lymphomas and Kaposi's sarcoma. A progressive, subacute encephalopathy is a frequent manifestation of this inevitably fatal syndrome. Autopsy studies have revealed numerous infectious causes of CNS disease, frequently multiple in the individual patient (MOSKOWITZ et al. 1984). Many of the CNS infections in patients with AIDS appear to be due to reactivation of a previously acquired agent. Thus toxoplasma encephalitis is exceptionally common and viral infections, particularly, progressive multifocal leucoencephalopathy and cytomegalovirus. In some patients, both adults and children, who have shown progressive neurological deterioration characterised by severe dementia, the post-mortem histological findings appear to be specific for AIDS (BUDKA 1986b; SHARER et al. 1986).

4.1.1 Pathological Changes of AIDS Encephalopathy

4.1.1.1 Specific Histological Changes of AIDS Encephalopathy

In these cases macroscopic appearances may be unremarkable, but usually there is obvious gyral atrophy, together with symmetrical enlargement of the ventricles. Severe microcephaly has been recorded in some paediatric patients (EPSTEIN et al. 1985). The cut surface of the deep cerebral white matter shows diffuse, roughly symmetrical greyish discolouration. The change may also be seen in the cerebellar white matter. Histology reveals widespread loss of myelin staining in deep white matter of all lobes of the cerebral hemispheres and in the cerebellum, but sparing subcortical arcuate fibres. Demyelinated areas contain macrophages and show reactive gliosis with numerous GFAP positive astrocytes. Axonal loss is less severe than myelin depletion. Small cellular clusters composed of lymphocytes, monocytes and occasional plasma cells, together with astrocytes are widely scattered in deep cerebral white matter, basal ganglia and brain stem. Sparse perivascular lymphocytes are an inconsistent finding. Vascular calcification, most commonly in the basal ganglia, is frequently observed in children. However, the most distinctive feature of AIDS encephalopathy is the presence of multinucleate giant cells within many of the cell aggregates (SHARER et al. 1986) and also dispersed throughout the white matter, particularly around blood vessels in demyelinated areas (KLEINHUES et al. 1985). These cells are quite different from the large, bizarre astrocytes of progressive multifocal leucoencephalopathy. The giant cells of AIDS encephalopathy have a rounded outline and a circle or semicircle of regular peripheral nuclei. They are not stained by the immunoperoxidase method for GFAP and are negative for all neuronal markers. Current evidence favours a monocyte-macrophage lineage. Morphologically these cells resemble Langhan's or Touton giant cells and routine histological staining characteristics, such as PAS positive, diastase resistant cytoplasmic granules, are similar to those of macrophages. Similar multinucleate cells have also been found in lymph nodes of AIDS patients. Immunocytochemical stains for macrophage markers have been inconclusive, but formalin-fixed post-mortem tissue is unsatisfactory for their demonstration (BUDKA 1986b).

4.1.1.2 Microglial Nodules in AIDS Encephalopathy

Small aggregates of mononuclear, microglial cells, unaccompanied by multinucleate giant cells and referred to as microglial nodules, have frequently been identified in the cerebral cortex of AIDS patients, especially adults. These foci are very probably related to CMV which infects neurons and occasionally typical 'owls eye' intranuclear inclusions can also be identified (NIELSEN et al. 1984). Microglial nodules are a frequent finding in immunosuppressed transplant patients, a high percentage of whom have been shown to have reactivation of CMV (BOOSS and ESIRI 1986).

4.1.2 Pathogenesis of AIDS Encephalopathy

HTLV-III is a retrovirus and has the morphology of a lentivirus. Lentiviruses are exogenous viruses which affect ungulates, causing persistent and debilitating infections, including encephalopathies. Visna virus, which causes demyelination in sheep, has a sequence homology with HTLV-III (Gonda et al. 1985). The lentiviruses infect cells of the immune system as well as the nervous system with resulting immune dis-regulation. The long association of man and domestic ungulates may be relevant to the appearance of the AIDS virus in man. In man T4 (helper) lymphocytes possess a surface receptor for HTLV-III and AIDS is characterised by a drastic reduction in numbers of T helper cells due to the cytopathic effects of the virus, hence the paucity of cell mediated immune response to any form of infection in AIDS. HTLV-III has been identified in bone marrow and there is limited productive infection in macrophages. Infected macrophages express I a antigen on the cell surface, induced by lentivirus interferon and viral replication is restricted by T cell – macrophage interaction (R.T. Johnson, personal communication). Despite the wide variety of pathogens that may cause CNS disease in AIDS patients it seems likely that the demyelinating encephalopathy with giant cells is due to CNS invasion by HTLV-III. The virus has been identified in brains of affected patients, both adults and children by in situ hybridisation (Shaw et al. 1985; Sharer et al. 1986) and by direct immunofluorescence using a monoclonal antibody to HTLV-III (Michaud et al. 1986). The immunohistochemical technique demonstrates virus in the CNS almost exclusively within mononuclear cells and in the multinucleate giant cells, both believed to be of haematogenous origin. Retroviral particles have been observed in the giant cells by electron microscopy (Sharer et al. 1986). Very few infected olgodendroglia, astrocytes or neurons have been detected by these techniques. It seems likely that the virus is carried to the CNS within blood monocytes, but there is seemingly a discrepancy between the small numbers of endogenous CNS cells that become infected and the extensive myelin breakdown. In young children reduction in white matter may be severe, because there is deficient myelin formation in addition to myelin loss. DNA hybridisation studies of brain tissue from one very young infant identified a far greater amount of HTLV-III genome than could be accounted for by the number of positive cells demonstrated by other techniques (Shaw et al. 1985). Thus, although CNS viral invasion occurs, the extent of infection of endogenous cells and the pathogenesis of AIDS encephalopathy is not fully elucidated.

4.2 Progressive Multifocal Leucoencephalopathy

Progressive multifocal leucoencephalopathy (PMLE) is an opportunistic infection of the CNS with a ubiquitous virus, that rarely has human pathogenecity unless there is a depression of cell mediated immunity. In the great majority of cases this virus is Papova virus, named J–C virus after the patient in whom it was first isolated. Two other closely related Papova viruses have rarely been identified, the simian virus SV40 and another isolated from one human patient and called BK virus. All three Papova viruses are oncogenic in small animals.

Until recent times PMLE was most often a complication of malignant lymphoma, where immunosuppression was due to the combined effects of disseminated lymphoid neoplasia and cytotoxic drugs employed in therapy. Today PMLE is also reported in a significant number of patients with AIDS (Moskowitz 1984). On rare occasions this CNS viral infection has developed in patients with no predisposing disorder. In the immunosuppressed patient PMLE is usually fatal within 6 months of clinical onset. Spontaneous remissions and prolonged survival of several years have occurred, particularly in those with no associated immunosuppressive illness and in others who have had at least temporary recovery of immune function. Dementia or some degree of mental deficit is a not infrequent clinical presentation, but major motor signs are also common and reflect the multiplicity of pathological lesions.

4.2.1 Pathological Changes

4.2.1.1 Macroscopic Appearances

PMLE is an accuratively descriptive term for a disorder in which multiple foci of myelin degeneration appear anywhere in the CNS white matter (Woodhouse et al. 1967). At post mortem external examination of the brain is usually entirely normal, except in the minority of unusually chronic cases, when reduced brain weight and cerebral atrophy may be evident. However in the cut surface of all cases many of the demyelinated foci will be visible to the naked eye, varying between tiny grey dots and irregular patches several cms in diameter (Fig. 10).

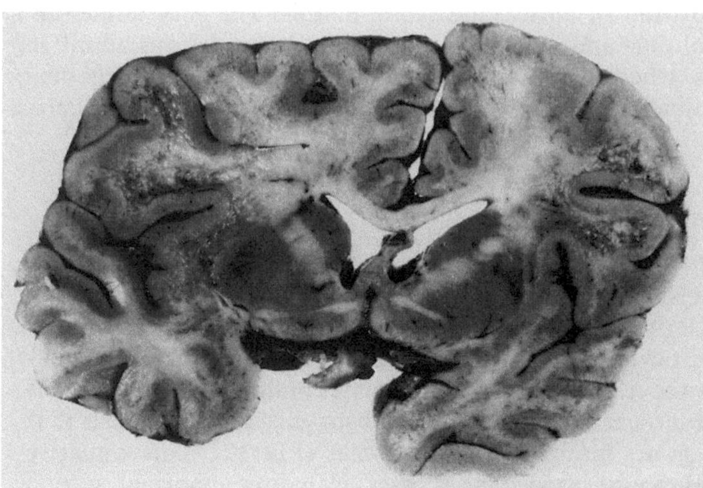

Fig. 10. Progressive multifocal leucoencephalopathy. Section of the brain from a 40 year old man with Hodgkin's disease, who died in coma 6 weeks after the onset of neurological symptoms. Extensive areas of greyish discolouration and softening in the white matter

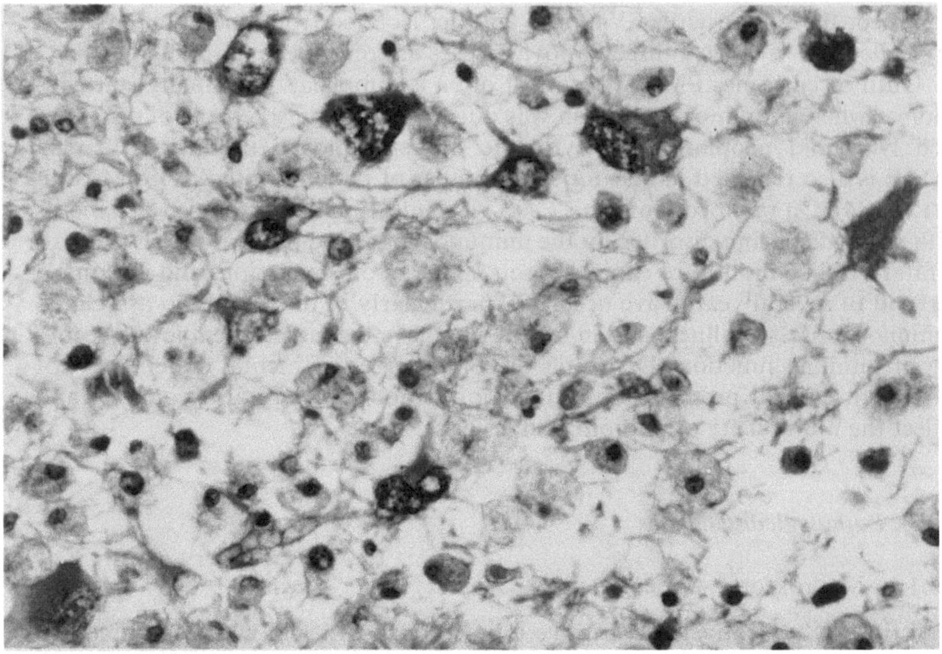

Fig. 11. Progressive multifocal leucoencephalopathy. A focus of myelin breakdown contains foamy macrophages and many large, bizarre astrocytes

Although no CNS white matter is exempt, lesions are usually most numerous in the cerebrum and the spinal cord is least involved. Multiple small lesions in the subcortical white matter may impart a motheaten appearance to the normally sharp demarcation of cortex and white matter. The very numerous tiny lesions, irregular outline and irregular scattering of larger lesions all differ from multiple sclerosis, where there is often rough symmetry, sharply circumscribed lesions one centimetre or more across and frequent periventricular distribution.

4.2.1.2 Histological Findings

The microscopy of PMLE is unique in the human CNS. In the typical case lesions show active myelin breakdown, with foamy macrophages engulfing myelin products, and partial preservation of axons. This demyelination is associated with swollen oligodendrocytes, reactive astrocytes and the most distinctive feature, many enlarged astrocytes with bizarre, sometimes multiple or multilobated nuclei (Fig. 11). The latter are morphologically indistinguishable from cells that can be found in high grade malignant astrocytomas (Fig. 12). The similarity is increased by the presence of occasional mitoses, including abnormal mitotic figures, Thus, a needle biopsy could be very difficult to interpret unless closely correlated with the clinical picture and appearances on computerised body-section radiography (CT). In contrast to PVE and multiple sclerosis the demye-

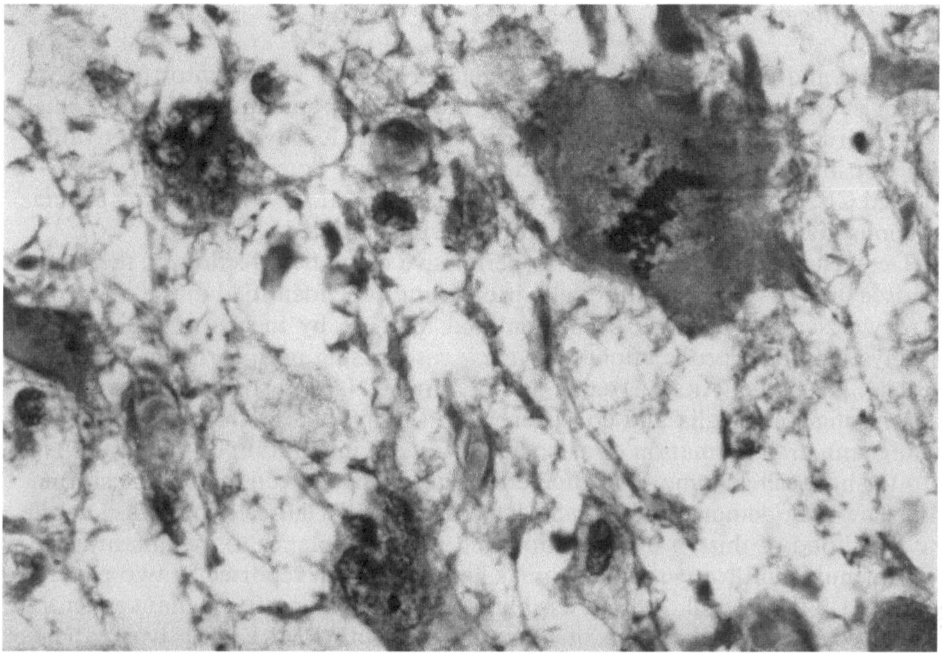

Fig. 12. Detail of the bizarre astrocytes in progressive multifocal leucoencephalopathy showing two multinucleate astrocytes and another large cell in mitosis

linating lesions of PMLE show no perivascular orientation. The lesions are generally devoid of any chronic inflammatory cell infiltrate, although sparse perivascular lymphocytes and plasma cells may be found in the protracted cases. As the lesion enlarges oligodendrocytes disappear from the centre, as do foamy macrophages when myelin debris has been removed, but the bizarre astrocytes persist.

4.2.1.3 Distribution of the Virus

Electron microscopy reveals Papova virus particles, spherical virions, often in crystalline arrays and filamentous forms in oligodendrocyte nuclei. Immunohistochemistry confirms that in most cases of PMLE these are J–C virus particles. The vast majority of cells that can be shown by these methods to contain virus are oligodendrocytes (Itoyama et al. 1982). Only a few, isolated oligodendroglia are stained by the immunoperoxidase technique in histological normal white matter, but where there is active demyelinated focus, many oligodendrocytes are labelled. The antigen is also demonstrable in oligodendrocyte cytoplasm and cell processes. Sparse reactive astrocytes are positive, but no viral antigen is demonstrated in the bizarre astrocytes with immunoperoxidase techniques. Oligodendrocytes and thus antigen containing cells are virtually absent from the centre of an established lesions when demyelination has taken place.

4.2.2 Pathogenesis

The route and time of entry of J–C or related Papova viruses into the nervous system is uncertain. A large percentage of the normal population have antibodies to J–C virus within the first two decades, thus PMLE may represent activation of latent infection, but where it resides is unknown. Some studies have shown infected CNS endothelial cells in PMLE (DORRIES et al. 1979) suggesting haematogenous dissemination, but this is not well substantiated.

Demyelination is attributable to permissive cytolytic infection of the oligodendrocyte by J–C virus. Although virions cannot be identified in the abnormal astrocytes, viral DNA has been identified in these by means of a J–C virus biotinylated DNA probe applied to a few cases, using both formalin-fixed and frozen CNS tissue (AKSAMIT et al. 1985). Viral DNA in astrocytes was far less than in oligodendroglia and consistent with the hypothesis that morphological 'malignant' transformation of astrocytes is due to integration of the viral DNA into the host cell genome. This dual effect, permissive infection and transformation, has been demonstrated by SV40 virus in fetal glial cell cultures. Despite the likelihood of this viral integration and the resemblance to malignant astrocytes, genuine infiltrative astrocytomas have only been reported in two patients with PMLE (SIMA et al. 1983). Possibly the full evolution of malignant potential is aborted by the early death of most patients with PMLE. The two patients who developed astrocytomas had unusually chronic disease, but other controlling host cell factors may operate. It is exciting to speculate that these Papova viruses may be responsible for spontaneous gliomas in man, but at present the links are tenuous and there is no firm evidence (JOHNSON 1982).

5 Unconventional or Slow Virus Infection

Recognition of the similarity between the unusual microscopic spongy appearance of the cerebral cortex in Kuru, a CNS degenerative disorder once common in New Guinea, and scrapie in sheep prompted the original transmission experiments which proved the infectious nature of Kuru (BENDHEIM 1984). The distinctive histology led further to the discovery of the close relationship between Kuru and the progressive dementia, Creutzfeldt-Jakob disease. Thus transmissible spongiform encephalopathy in man was promoted from a pathological curiosity of remote New Guinea to a disease with global distribution (MATTHEWS 1986).

The incidence of Creutzfeldt-Jakob disease is estimated to be approximately 200 new cases per year in the USA. Up to 15% of cases appear to be familial, but this could be due either to inherited susceptibility or common exposure to an ennvironmental pathogen, BENDHEIM (1984). Most cases are sporadic and isolated and the disease does not seem to be readily communicable. Despite laboratory evidence of a transmissable pathogen the mechanism of natural transmission in man is unknown. Occasional small clusters have been attributed to iatrogenic transmission, such as occurred through use of contaminated neurosurgical instruments (BERNOULLI et al. 1979). Accidental transmission has also occurred with injections of pituitary growth hormone obtained from human cadavers (WELLER et al. 1986) and from human dura (Morbidity and Mortality Weekly Report CDC 1986). The disease most often affects the late middle-aged and elderly, although the young are not exempt. Typically a psychiatric disturbance heralds a progressive dementia accompanied by myoclonus and associated with characteristic EEG abnormalities. Terminally the patient becomes decerebrate and comatose and the whole course rarely exceeds one year. Clinical variants include cases with prominent cerebellar signs (BOOSS and ESIRI 1986).

The exact nature of the infectious agents is still uncertain, but unique properties distinguish them from conventional viruses. The histopathological changes of the encephalopathy are now well described and sufficiently unusual to be virtually diagnostic at post-mortem (MASTERS 1978; BOOSS and ESIRI 1986) although complete proof demands successful animal transmission. Histology and electron microscopy have shed very little light upon the nature of the infectious agents, but this has been vigorously pursued through biochemical and biophysical studies.

5.1 Pathological Changes

5.1.1 Macroscopic Appearances and Histology

At autopsy, despite the clinical neurological devastation, the brain may have an entirely normal macroscopic appearance. Cases with an unusually prolonged clinical course show cerebral cortical atrophy and cerebellar atrophy is sometimes striking in patients whose clinical history has been dominated by cerebellar signs (BOOSS and ESIRI 1986). Microscopical abnormalities in Creutzfeldt-Jakob disease predominate in grey matter. The cerebral and cerebellar cortices, basal ganglia and thalamus may all be involved to a variable degree. The picture, which is practically pathognomonic, is a very fine vacuolation of the neuropil, hence the term "spongiform" encephalopathy (Fig. 13). The fine vacuoles are quite separate from cell bodies. Initially, the vacuolar change may be most conspicuous in layers 2 and 3 of the cortex, but when advanced creates a delicate sieve like appearance of the full thickness of the cortex. Vacuolation is accompanied by loss of neurons and by astrocytic hypertrophy and hyperplasia. The number of astrocytes and degree of hyperplasia can only be appreciated by using special stains. Cajal's time honoured gold-chloride method (Fig. 14) and immunoperoxidase method for glial fibrillary acidic protein (GFAP) are both valuable. Astrocytic gliosis ultimately involves all cortical layers, but is particularly dense in the deep layers 4–6. Neuronal loss may be extensive in severely gliotic areas. Paradoxically the spongiosis may be less conspicuous in advanced disease when neuronal loss is greatest (WELLER et al. 1986) probably because it is then obscured by the intense gliosis. An important negative feature of Creutzfeldt-Jakob disease is the absence of any host inflammatory cell response. Transmissible spongiform encephalopathy must not be confused histologically with the common fixation artefact of perineuronal vacuolation or the coarse

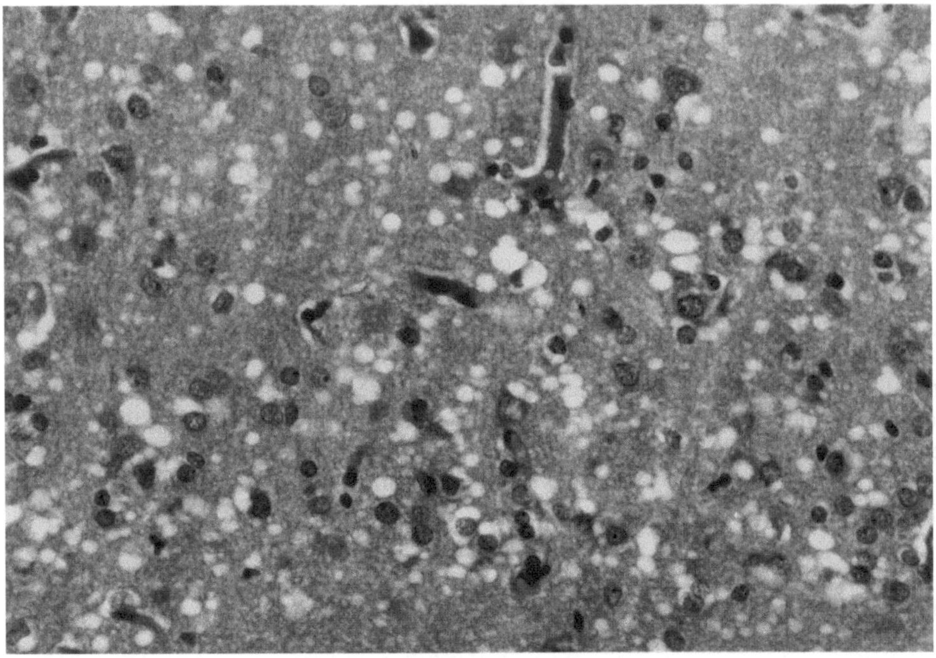

Fig. 13. Creutzfeldt-Jakob disease in a 64 year old man with progressive dementia and myoclonus. Section of cerebral cortex showing fine vacuolation of the neuropil

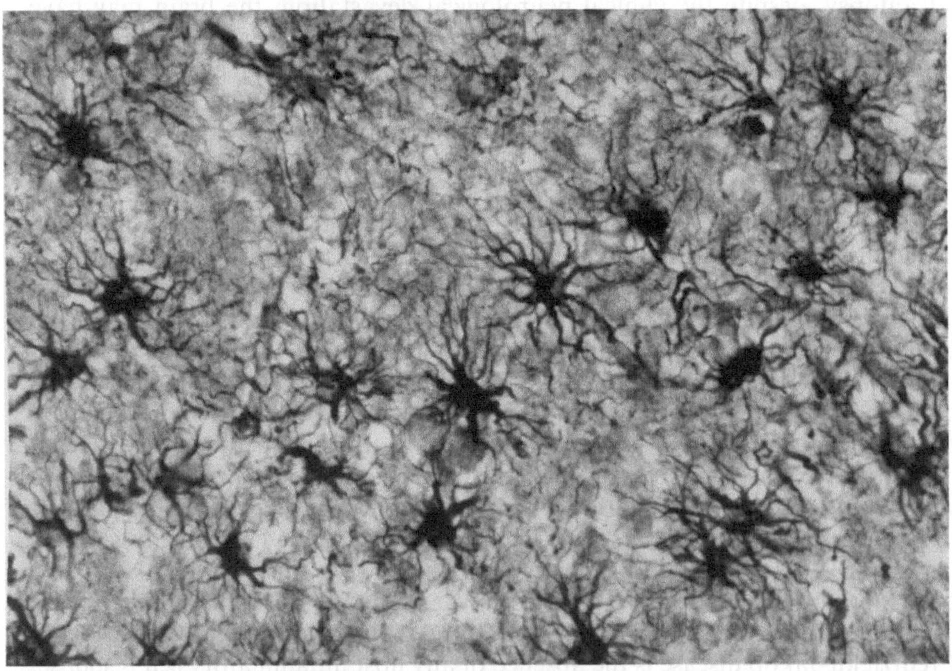

Fig. 14. Creutzfeldt-Jakob disease. Same case as in Figure 16. Numerous hypertrophic astrocytes in the deeper layers of the cerebral cortex. (Cajal's gold-chloride stain)

irregular cavitation that is a late sequel to extensive neuron loss from any cause, such as profound hypoxia. The key feature of Creutzfeldt-Jakob disease is the very fine vacuolation of the neuropil at a stage when many neurons are still present and there is comparatively mild gliosis. Even in very chronic cases some areas of the cortex will have this appearance.

5.1.2 Electron Microscopy

Electron microscopy has helped to elucidate the spongiform change, but has not revealed the infectious agent. Spiroplasmas (cell membrane associated mycoplasma like organisms) were briefly incriminated (Bastian 1979). Membrane bound spiral elongated bodies observed in Creutzfeldt-Jakob disease and originally interpreted as spiroplasmas have since been observed in the normal developing rodent brain and it is suggested that their appearance in the degenerate cortex is due to abortive neurite regeneration (Gray 1986). The characteristic spongy appearance is largely due to ballooning of neuronal dendrites, very rarely swollen neuronal perikarya and a minor contribution from swollen astrocytic processes. Dendrites show depletion or even complete loss of microtubules. Synaptic contacts are sometimes recognisable on vacuoles devoid of cytoplasmic organelles. Dendrites are also reported to contain abnormal 6 nm filaments. The hypothesis based on these findings is that in Creutzfeldt-Jakob disease and the other unconventional viral infections, early breakdown of the dendritic microtubule skeleton inhibits normal dendritic function and post-synaptic events (Gray 1986).

5.2 The Putative Infectious Agent

The nature of the infectious agent is still a matter of controversy. Research efforts have been concentrated on the closely related scrapie agent. The incubation period for scrapie in a particular animal host is predictable and determined by the host genome (Kimberlin 1986). A single host gene controls the overall dynamics of scrapie replication in vivo. Although the scrapie agent can replicate in the lymphoreticular system, pathogenetic effects are only observed after CNS invasion. Experimental studies suggest that the agent spreads intraaxonally within the CNS, although at a much slower rate than a conventional virus such as Herpes simplex (Kimberlin 1986; Fraser et al. 1986).

A recent contender for the infectious agent in scrapie and the human transmissable encephalopathies is the PROIN, a term introduced by Prusiner et al. (1982) to denote a small PROteinaceous, INfectious particle resistant to procedures that denature and inactivate nucleic acids. Investigation of subcellular fractions of scrapie brain indicate that infectivity resides in the membrane fraction. Analyses of purified extracts of high infectivity have revealed that a protein (molecular weight 27000–30000) is the single, major constituent. This protein molecule has not been separated from scrapie infectivity, hence the introduction of the term prion and the protein has been designated PrP 27–30. Elimination of PrP from the purified extract destroys its infectivity, implying that the protein is not only essential for infectivity, but is also a component of the scrapie agent (PRUSINER et al. 1985). An alternative view is that a small scrapie agent is in some way attached to this protein and has yet to be separated.

Nucleic acid encoding PrP 27–30 cannot be extracted from the infectious particles and all evidence suggests the protein is the product of a cellular gene. Amino acid sequencing of PrP 27–30 was followed by reconstruction of the nucleotide sequence of DNA encoding PrP 27–30. A cloned DNA probe established that the prion protein PrP 27–30 is encoded by a single cellular gene in the hamster and homologous gene sequences are present in human and murine DNA (PRUSINER et al. 1985). PrP messenger RNA is present in similar amounts in both infected and normal brain cells, principally neurons. Levels of PrP mRNA apparently do not change during the course of infection although PrP 27–30 accumulates in diseased brain, whilst a very closely related, antigenically similar protein is present in lower levels in normal brain cells and appears to be incorporated into the cell membrane. PrP 27–30 has properties which distinguish it from the normal cell protein, but it is not known whether the differences are due to a minor mutation or some post-translational modification of the molecule. PrP protein in scrapie brain shows greater resistance to proteinase digestion and can be induced to polymerise into rods or filaments, visible by electron microscopy. These structures, referred to as scrapie associated fibrils (SAF) have been isolated not only from different scrapie sources, but also from Creutzfeldt-Jakob disease and Kuru brain extracts (KIMBERLIN 1986). The fibrils are composed of 2–4 twisted filaments, each 4–6 nm diameter. The histopathology of scrapie is very similar to Creutzfeldt-Jakob disease, comprising grey matter vacuolation, astrocytic gliosis and neuronal loss, but in addition extra cellular deposits with the tinctorial properties of amyloid are a frequent finding (GILMOUR et al. 1986). Immunohistochemical methods using monoclonal antibodies raised to PrP 27–30 applied to sections of scrapie brain selectively stain the amyloid fibrils and no other structures (PRUSINER 1982). Amyloid plaques are not as conspicuous in Creutzfeldt-Jakob disease as in scrapie, but may occur and these are likewise stained with PrP antibodies. Other forms of human cerebral amyloid, including Alzheimer plaques are unstained (KIMBERLIN 1986).

Thus molecular probes have confirmed the links between scrapie, Kuru and Creutzfeldt-Jakob disease that were suspected from the original observations of histological similarities. However the exact nature of the infectious agent remains an enigma. A modified normal cell membrane protein that is sometimes deposited extracellularly in the diseased tissue, is apparently an essential component, but the biological variation between different scrapie strains and ability of the agent to replicate suggest there must also be a nucleic acid component. The resistance of the scrapie agent to standard procedures that denature nucleic acid is well known and clearly the agent does not contain the nucleic acid that encodes its protein. Nevertheless this does not exclude the possibility of another very small nucleic acid molecule bound to the protein. A recent analysis of purified cell membrane derived vesicles from scrapie infected hamster brain revealed the presence of low molecular weight RNA, even after nuclease treatment (DEES et al. 1985). Although conventional viruses redirect host cell metabolism to manufacture their protein constituents, the code is supplied by the virus.

An agent coding only for limited biological properties and obtaining the protein required for infectivity entirely from the host genome would be unique. It would however explain the complete lack of antigenicity of the transmissable agent. The term "virino" has been proposed (KIMBERLIN 1986).

Although laboratory experiments fulfil the criteria of infectious disease, both Creutzfeldt-Jakob disease and scrapie may show patterns of inherited disease. The absence of communicability in man has been mentioned. The possibility that the whole agent is normally endogenously derived must therefore be considered. The answers to such speculation will come from identification of any hidden informational molecule and explanation of the differences between the scrapie protein and its normal cell counterpart. The infectious agent in some way disrupts the metabolism of neurons, causing their degeneration and astrocytic gliosis has usually been considered as a non specific response. However immortalised, transformed cell lines can be obtained from cultures of scrapie infected brain that include both glial and mesodermal elements. Transplantation of these cells into nude mice has produced neoplasms (MANUELIDIS 1985). Astrocytic gliosis is an early change in Creutzfeldt-Jakob disease and it is conceivable that these cells are being induced to proliferate by the agent. If so it provides further support for a nucleic acid and even suggests this could be a form of oncogene.

References

Aaby P, Bukh J, Lisse IM, Smits AJ (1984) Risk dependent factors in subacute sclerosing panencephalitis: Age- and sex-dependent host reactions or intensive exposure? Rev Infect Dis 6:239–250

Anderson JR, Field HJ (1983) The distribution of herpes simplex type 1 antigen in mouse central nervous system after different routes of inoculation. J Neurol Sci 60:181–195

Barber PC, Lindsay RM (1982) Schwann cells of the olfactory nerves contain glial fibrillary acidic protein and resemble astrocytes. Neuroscience 7:3077–3090

Bastian FO, Wende RD, Singer DB, Zeller RS (1975) Eastern equine encephalitis. Histopathologic and ultrastructural changes with isolation of the virus in a human case. Am J Clin Pathol 64:10–13

Bastian FO (1979) Spiroplasma-like inclusions in Creutzfeldt-Jakob disease. Arch Pathol Lab Med 103:665–669

Benator RM, Magill HL, Gerald B, Igarashi M, Fitch SJ (1985) Herpes simplex encephalitis: CT findings in the neonate and young infant. ANJR 6:539–543

Bendheim PE (1984) The human spongiform encephalopathies. In: Johnson KP (ed) Symposium on neurovirology. Saunders, Philadelphia, pp 281–298 (Neurologic Clinics, vol 2)

Bernouilli CC, Masters CL, Gajdusek DC (1979) Early clinical features of Creutzfeldt-Jakob disease. In: Slow transmissable diseases of the nervous system, vol 1. Academic Press, New York

Booss J, Esiri MM (1986) Viral encephalitis. Pathology, diagnosis and management. Blackwell Scientific Publications, Oxford

Britton CB, Mesa-Tejada R, Fenoglio CM, Hays AP, Garvey GE, Miller JR (1985) A new complication of AIDS: Thoracic myelitis caused by herpes simplex virus. Neurology 35:1071–1074

Budka H (1986a) Selective vulnerability of specific cell populations in human CNS infections. X International Congress of Neuropathology, Stockholm (abstract)

Budka H (1986b) Multinucleated giant cells in the brain: A hallmark of the acquired immune deficiency syndrome (AIDS). Acta Neuropathol (Berl) 69:253–258

Choppin PW (1981) Measles virus and chronic neurological diseases. Ann Neurol 9:17–20

Dayan AD, Gooddy W, Harrison MJG, Rudge P (1972) Brain stem encephalitis caused by herpesvirus hominis. Br Med J 4:405–406

Dees C, McMillan BC, Wade WF, German TL, Marsh RF (1985) Characterisation of nucleic acids in membrane vesicles from scrapie-infected brain. J Virol 55:126–132

Dix RD, Waitzman DM, Follansbee S, Pearson BS, Mendelson T, Smith P, Davis RL, Mills J (1985) Herpes simplex virus type 2 encephalopathy in two homosexual men with persistent lymphadenopathy. Ann Neurol 17:203–206

Efstathiou S, Minson AC, Field HJ, Anderson JR, Wildy PW (1986) Detection of herpes simplex virus-specific DNA sequences in latantly affected mice and in humans. J Virol 57:446–455

Embil JA, Camfield PR, Artsob H (1985) California encephalitis in New Brunswick. Can Med Assoc J 132:1166

Epstein LG, Sharer LR, Joshi VV, Fojas MM, Koenigsberger MR, Oleske JM (1985) Progressive encephalopathy in children with acquired immune deficiency syndrome. Ann Neurol 17:488–496

Esiri MM, Oppenheimer DR, Brownell B, Haire M (1981) Distribution of measles antigen and immunoglobulin containing cells in the CNS in subacute sclerosing panencephalitis (SSPE) and atypical measles. J Neurol Sci 53:29–43

Esiri MM (1982) Herpes simplex encephalitis. An immunohistological study of the distribution of viral antigen within the brain. J Neurol Sci 54:209–226

Field HJ, Anderson JR, Efstathiou S (1984) A quantitative study of the effects of several nucleoside analogues on established herpes simplex encephalitis in mice. J Gen Virol 65:707–719

Fournier J-G, Tardieu M, Lebon P, Robain O, Ponsot G, Rozenblatt S, Bouteille M (1985) Detection of measles virus RNA in lymphocytes from peripheral-blood and brain perivascular infiltrates of patients with subacute sclerosing panencephalitis. N Engl J Med 313:9910–9915

Gamboa ET, Cowen D, Eggers MD, Cogan P, Ganti R, Brust JCM (1983) Delayed onset of post-rabies vaccination encephalitis. Ann Neurol 13:676–678

Gatus BJ, Rose MR (1983) Japanese B encephalitis: epidemiological, clinical and pathological aspects. J Infect 6:213–318

Gendelman HE, Wolinsky JS, Johnson RT, Pressman NJ, Pezeshkpour GH, Boisset GF (1984) Measles encephalomyelitis: Lack of evidence of viral invasion of the central nervous system and quantitative study of the nature of demyelination. Ann Neurol 15:353–360

Gilden DH, Rorke LB, Tanaka R (1975) Acute SSPE. Arch Neurol 32:644–646

Gilmour JS, Bruce ME, MacKellar A (1986) Neuropathol Appl Neurobiol 12:173–183

Gonda MA, Wong-Staal F, Gallo RC, Clements JE, Narayan O, Gilden RV (1985) Sequence homology and similarity of HTLV-III and a visna virus, a pathogenic lenivirus. Science 227:173–177

Gonzalez-Scarano F (1986) Molecular basis of neuroinvasiveness and neurovirulence. X International Congress of Neuropathology, Stockholm (abstract)

Gray EG (1986) Spongiform encephalopathy: a neurocytologist's viewpoint with a note on Alzheimer's disease. Neuropathol Appl Neurobiol 12:149–172

Graves MC (1984) Subacute sclerosing panencephalitis. In: Johnson KP (ed) Symposium on neurovirology. Saunders, Philadelphia, pp 267–280 (Neurologic Clinics, vol 2)

Greenwood R, Bhalla A, Gordon A, Roberts J (1983) Behaviour disturbances during recovery from herpes simplex encephalitis. J Neurol Neurosurg Psychiat 46:809–817

Grimley PM (1983) Arbovirus encephalitis: which road travelled makes all the difference. Lab Invest 48:369–371

Hirsch RL, Griffin DE, Johnson RT, Cooper SJ, De Sobriano IL, Roedenbeck S, Vaisberg A (1984) Clin Immunol Immunopathol 31:1–12

Ishii T, Matsushita M, Hamada S (1977) Characteristic residual neuropathological features of Japanese B encephalitis. Acta Neuropathol (Berl) 38:181–186

Itoyama Y, Webster H de F, Sternberger NH, Richardson EP, Walker DL, Quarles RH, Padgett BL (1982) Distribution of papova virus, myelin-associated glycoprotein, and myelin basic protein in progressive multifocal leucoencephalopathy lesions. Ann Neurol 11:396–407

Jenista JA (1983) Perinatal herpes virus infections. Semin Perinatol 7:9–15

Johnson RT (1964) The pathogenesis of herpes simplex encephalitis – Virus pathways to the nervous system of suckling mice demonstrated by fluorescent antibody staining. J Exp Med 119: 343–356

Johnson RT (1984) Viral infections of the nervous system. Raven Press, New York

Johnson RT, Griffin DE, Hirsch RL, Wolinsky JS, Roedenbeck S, De Sobriano IL, Vaisberg (1984) Measles encephalitis – clinical and immunologic studies. N Engl J Med 310:137–141

Johnson RT, Burke DS, Elwell M, Leake CJ, Nisalak A, Hoke CH, Lorsomrudee W (1985) Japanese

encephalitis: immunocytochemical studies of viral antigen and inflammatory cells in fatal cases. Ann Neurol 18:567–573
Kimberlin RH (1982) Reflection on the nature of the scrapie agent. Trends Biochem Sci 7:392–394
Kimberlin RH (1986) Scrapie: how much do we really understand? Neuropathol Appl Neurobiol 12:131–147
Kohl S, James AJ (1985) Herpes simplex encephalitis during childhood: Importance of brain biopsy diagnosis. J Pediatr 107:212–215
Koskininiemi M, Vaheri A, Taskinen E (1984) Cerebrospinal fluid alterations in herpes simplex encephalitis. Rev Infect Dis 6:608–618
Knobbler RL, Haspel MV, Oldstone MBA (1981) Mouse hepatitis virus type-4 (JHM) strain-induced fatal central nervous system disease. 1. Genetic control and the murine neuron as the susceptible site of disease. J Exp Med 153:832–43
Kristensson K, Nennesmo I, Persson L, Lycke E (1982) Neuron to neuron spread of herpes simplex virus. Transport of virus from skin to brainstem nuclei. J Neurol Sci 54:149–56
Land SA, Skurrie IJ, Gilbert GL (1984) Rapid diagnosis of herpes simplex encephalitis infections by enzyme-linked immunosorbent assay. J Clin Microbiol 19:865–869
Longson M, Bailey AS (1980) Herpes simplex encephalitis. In: Waterson AP (ed) Recent advances in virology, vol 1. Churchill Livingstone, London, pp 1–21
Luby JP (1979) St. Louis encephalitis. Epidemiol Rev 1:55–73
Malamud N, Haymaker W, Pinkerton H (1950) Inclusion encephalitis: with a clinicopathological report of 3 cases. Am J Pathol 26:133–153
Manuelidis EE (1985) Creutzfeldt-Jakob disease. J Neuropath Exp Neurol 44:1–17
Martin JR, Nathansan N (1979) Animal moels of virus induced demyelination. In: Zimmerman H (ed) Progress in neuropathology, vol 4. Raven Press, New York, pp 27–50
Masters CL, Richardson EP (1978) Subacute spongiform encephalopathy (Creutzfeldt-Jakob disease). The nature and progression of the spongiform change. Brain 101:333–344
Matthews WB (1986) Unconventional virus infection and neurologic disease. Neuropathol Appl Neurobiol 12:111–116
Michaud J, Pekovic D, Lapointe N, Geoffrey G, Ajdukovic D (1986) Detection of HTLV-III/LAV in brains of two pediatric cases of AIDS-related subacute encephalitis. X International Congress of Neuropathology, Stockholm (abstract)
Minkler DS, McLean EB, Shaw CM, Hendrickson A (1976) Herpes virus hominis encephalitis and retinitis. Arch Ophthalmol 94:89–95
Monath TP, Cropp CB, Harrison AK (1983) Mode of entry of a neurotropic arbovirus into the central nervous system. Lab Invest 48:399–410
Morbidity and Mortality Weekly Report, Communicable Disease Centre, Atlanta (1986) 36(4):49–50
Moskowitz LB, Hensley GT, Chan JC, Gregorios J, Conley FK (1984) The neuropathology of the acquired immune deficiency syndrome. Arch Pathol Lab Med 108:867–872
Nahmias AJ, Whitley RJ, Visintine AN, Takei Y, Alford CA (1982) Herpes simplex encephalitis: Laboratory evaluations and their diagnostic significance. J Infect Dis 145:829–836
Nielsen SL, Petito CK, Urmacher CD (1984) Subacute encephalitis in acquired immune deficiency syndrome: a post-mortem study. Am J Clin Pathol 82:678
Oesch B, Westaway D, Walchli M, Mckinley MP, Kent SBH, Aebersold R, Barry RA, Tempst P, Teplow DB, Hood LE, Prusiner SB, Weissman C (1985) A cellular gene encodes scrapie PrP 27–30 protein. Cell 40:735–746
Ojeda VJ, Archer M, Robertson TA, Bucens MR (1983) Necropsy study of the olfactory portal of entry in herpes simplex encephalitis. Med J Aust 1:79–81
Paula-Barbosa MM, Brito R, Silva CA, Faria R, Cruz C (1979) Neurofibrillary changes in cerebral cortex of a patient with subacute sclerosing panencephalitis (SSPE). Acta Neuropathol (Berl) 48:157–160
Paula-Barbosa MM, Tavares MA, Saraiva AA (1980) Dendritic abnormalities in patients with SSPE. A golgi study. Acta Neuropathol (Berl) 52:77–80
Paula-Barbosa MM, Tavares MA, Borges MM (1984) Mitochondrial abnormalities in cortical dendrites from patients with early forms of subacute sclerosing panencephalitis (SSPE). Acta Neuropathol (Berl) 63:117–122
Phinney PR, Fliegiel S, Bryson YJ, Porter DD (1982) Necrotising vasculitis in a case of disseminated neonatal herpes simplex infection. Arch Pathol Lab Med 106:64–67

Prusiner SB (1982) Novel proteinaceous infectious particles cause scrapie. Science 216:136–144
Rennick PM, Nolan DC, Bauer RB, Lerner AM (1973) Neuropsychologic and neurologic follow up after herpes virus hominis encephalitis. Neurology 23:42–47
Risk WS, Haddad FS, Chemali R (1978) Substantial spontaneous long term improvement in subacute sclerosing panencephalitis. Arch Neurol 35:494–502
Rivers TM, Swentker FF (1935) Encephalomyelitis accompanied by myelin destruction experimentally produced in monkeys. J Exp Med 61:698–702
Roman-Campos G, Toro G (1980) Herpetic brainstem encephalitis. Neurology 30:981–985
Ryder JW, Croen K, Kleinschmidt-De-Masters BK, Ostrove JM, Straus SE, Cohn DL (1986) Progressive encephalitis three months after resolution of cutaneous zoster in a patient with AIDS. Ann Neurol 19:182–188
Sever JL (1983) Persistent measles infection of the central nervous system: Subacute sclerosing panencephalitis. Rev Infect Dis 5:467–473
Shankar SK, Rao TV, Mruthyunjayanna BP, Devi MG, Deshpande DH (1983) Autopsy study of brains during an epidemic of Japanese encephalitis in Karnataka. Indian J Med Res 78:431–440
Sharer LR, Epstein LG, Cho E-S, Joshi VV, Meyenhofer MF, Rankin LF, Petito CK (1986) Pathologic features of AIDS encephalopathy in children. Evidence for LAV/HTLV-III infection of brain. Hum Pathol 17:271–284
Shaw GM, Harper ME, Hahn BH, Epstein LG, Gajdusek DC, Price RW, Navia BA, Petito CK, O'Hara CJ, Groopman JE, Cho E-S, Oleske JM, Wong-Staal F, Gallo RC (1985) HTLV-III infection in brains of children and adults with AIDS encephalopathy. Science 227:177–181
Silva CA, Paula-Barbosa MM, Peteia S, Cruz C (1981) Two cases of rapidly progressive subacute panencephalitis. Neuropathological findings. Arch Neurol 38:109–113
Sima AAF, Finkelstein SD, McLachlan DR (1983) Multiple malignant astrocytomas in a patient with spontaneous progressive multifocal leucoencephalopathy. Ann Neurol 14:183–188
Skoldenberg B et al. (1984) Acyclovir versus vidarabine in herpes simplex encephalitis Lancet 1:707–711
Toro G, Vergara I, Roman G (1977) Neuroparalytic accidents of antirabies vaccination with suckling mouse brain vaccine. Clinical and pathologic study of 21 cases. Arch Neurol 34:694–700
Twomey JA, Barker CM, Robinson G, Howell DA (1979) Olfactory mucosa in herpes simplex encephalitis. J Neurol Neurosurg Psychiat 42:983–987
Tyler KL, Bronson RT, Byers KB, Fields B (1985) Molecular basis of viral neurotropism: experimental reovirus infection. Neurology 35:88–92
Weller RO, Steart PV, Powell-Jackson JD (1986) Pathology of Creutzfeldt-Jakob disease associated with pituitary-derived human growth hormone administration. Neuropathol Appl Neurobiol 12:117–129
Whitley RJ, Alford CA (1982) Herpes infections in childhood: diagnostic dilemmas and therapy. Pediatric Infect Dis 1:81–84
Whitley RJ, Soong SJ, Linneman C, Liu C, Pazin G, Alford CA (1982a) Herpes simplex encephalitis. Clinical assessment. JAMA 247:317–320
Whitley RJ, Lakeman AD, Nahmias A, Roizman B (1982b) DNA restriction-enzyme analysis of herpes simplex virus isolates obtained from patients with encephalitis. N Engl J Med 307:1060–1062
Whitley RJ (1984a) Treatment of human herpes virus infections with special reference to encephalitis. J Antimicrobial Chemother 14 (Suppl A):57–74
Whitley RJ (1984b) Infections due to herpes simplex virus in the immunocompromised host. J Infect Dis 150:323–329
Woodhouse MA, Dayan AD, Burston J, Caldwell I, Hume Adams J, Melcher D, Ulrich H (1967) Progressive multifocal leukoencephalopathy: electron microscope study of four cases. Brain 90:863–870

A General Approach to Neuropathological Problems

R.O. WELLER

1 Introduction

There is a variety of different reasons for performing neuropathological investigations, but the general approach in each case is very similar. Examination of biopsy or autopsy material may be required to establish a diagnosis for the treatment of a patient or for clinico-pathological correlation. In other cases, information derived from neuropathological studies may be used for medicolegal purposes or for genetic counselling. The examination of biopsies from central nervous system lesions, muscle and peripheral nerve has the immediacy of a clinical diagnosis for the treatment of a patient whereas the documentation of neuropathology at post mortem has not only a diagnostic role but also plays an important part in the continuing education of neurologists, neurosurgeons and pathologists. For example, the examination of cerebral tumours at autopsy enables the pathologist to obtain a complete view of lesions which he usually sees only as small, fragmented biopsies.

Perhaps more than in any other organ system, the information derived from the clinical history, examination and investigation of a neurological patient is essential for localising the pathological lesion and for indicating the pathological nature of the disease. Recent advances in imaging techniques, particularly computerised axial tomography (CT scanning) and Magnetic Resonance Imaging (MRI) have revolutionised clinical neurology. However, it has become apparent that even these techniques cannot completely surplant the certainty and accuracy of a pathological diagnosis.

For the pathologist, the anatomical and histological complexity of the central nervous system presents a challenge which can only be met by a logical approach to the pathological examination of the brain, spinal cord, peripheral nerves and muscle. This chapter will concentrate first, upon how to derive the relevant information from the patient's clinical history and examination so that the appropriate neuropathological investigations can be planned and second, upon how to derive information from neuropathological material to enable the pathologist to reach a logical conclusion and diagnosis. Techniques for the examination of a fixed brain will be outlined and the methods by which data can be obtained from macroscopic appearances and from the judicious selection of histological material will be reviewed. Often, one of the most difficult problems in neuropathology is the detection and evaluation of histological changes. But, once these changes have been observed and their anatomical pattern documented, it is relatively easy to determine the diagnosis and pathological implications from accounts in the literature and from previous experience. The plethora of different histological stains applicable to the nervous system is a reflection of the difficulty in analysing the complex cellular inter-relationships particularly within the brain and spinal cord. Although to some extent the prospect of employing a large number of stains may seem daunting, their logical use simplifies the histological analysis of the tissue and brings more certainty to the identification of pathological changes. Many of the major histological features may be seen in Haematoxylin and eosin stained sections, and this technique is particularly suitable for the analysis of nuclear changes in tumours. However, many of the subtle changes in neuropathological lesions require a view of the whole cell and this may only be possible by the demonstration of the cytoplasmic processes which are not visible in haematoxylin and eosin stained sections. The use, therefore, of selected neuropathological techniques will often greatly simplify the interpretation of the histological changes.

2 Clinical Features as a Guide to Neuropathology

A considerable amount of time and effort may be saved if the clinical history and examination of the patient are evaluated before an autopsy is planned or a biopsy is examined.

The position of a pathological lesion is often very accurately localised by the clinical neurologist from his knowledge of neuroanatomy and clinical neurophysiology. As a straightforward example, a stroke causing a hemiparesis is usually due to an infarct or haemorrhage involving the internal capsule in one cerebral hemisphere, whereas paresis on both sides of the body usually means that the lesion is in the pons, medulla or spinal cord where the cortico-spinal tracts lie in close proximity. Intrinsic brain stem lesions can be localised clinically by the pattern of involvement of long tracts and cranial nerves whereas the clinical features of a truncal ataxia, intention tremor or nystagmus may suggest a cerebellar lesion. There are various features in the clinical history which may be valuable for narrowing the differential diagnosis of a pathological lesion (WALTON 1985; WELLER et al. 1983). It is important to know the patient's age as a number of diseases, such as brain damage due to birth hypoxia and posterior fossa tumors, usually present in childhood, but other disease entities, such as strokes and tumours in the cerebral hemispheres, occur more frequently in older patients. The clinical progression of a lesion may be a valuable clue to its pathology. A sudden onset of unconsciousness or hemiparesis usually suggests a stroke due to an infarct or haemorrhage whereas slowly progressive paralysis of one or more limbs would be more typical of a tumour. Many neurological diseases have a genetic basis (BUNDEY 1985), so that the family history is often a key to the diagnosis particularly in diseases such as Huntingdon's chorea, lipidoses and Duchenne muscular dystrophy. Cytogenetic investigation of chromosomal abnormalities, particularly in mentally retarded children, may be a valuable guide to the pattern of expected neuropathology. In the future, DNA probes may enhance the accuracy of the clinical diagnosis of Huntingdon's chorea and Duchenne dystrophy (EMERY 1985). Already, the biochemical demonstration of enzyme defects in blood leucocytes or in fibroblasts has largely surplanted the need for biopsy diagnosis of inborn errors of lysosomal enzyme metabolism such as the gangliosidoses and the glycogenoses. Nevertheless, clinico-pathological correlation in these diseases is still required to determine the pattern of damage in individual cases and such pathological evaluation will be increasingly important in evaluating any methods of treatment (LAKE 1984).

3 Imaging Techniques and Neuropathology

The introduction of computerised axial tomography (CT scanning), and more recently Magnetic Resonance Imaging (Nuclear Magnetic Resonance (NMR)) and Positron Emission Tomography have greatly improved imaging of the brain and spinal cord (GONZALEZ et al. 1985). Older radiological techniques such as angiography continue in their sophistication and especially with the use of digital subtraction angiography which has reduced the need for large volumes

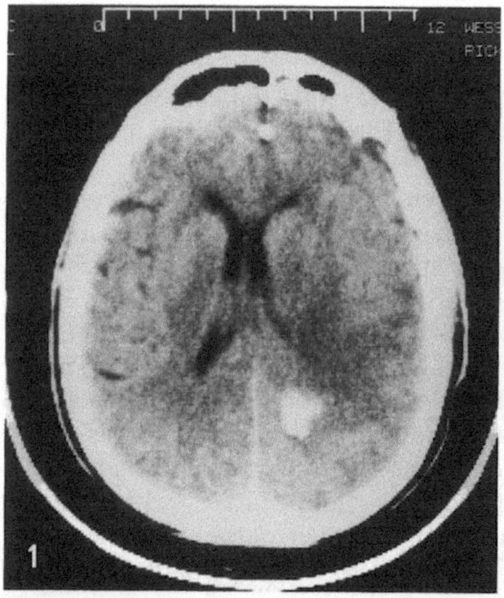

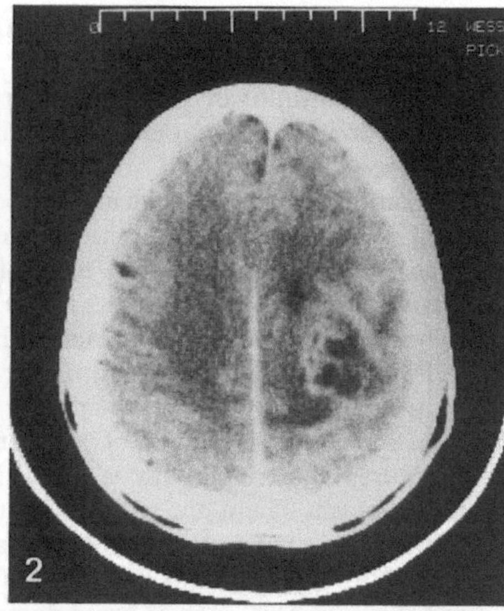

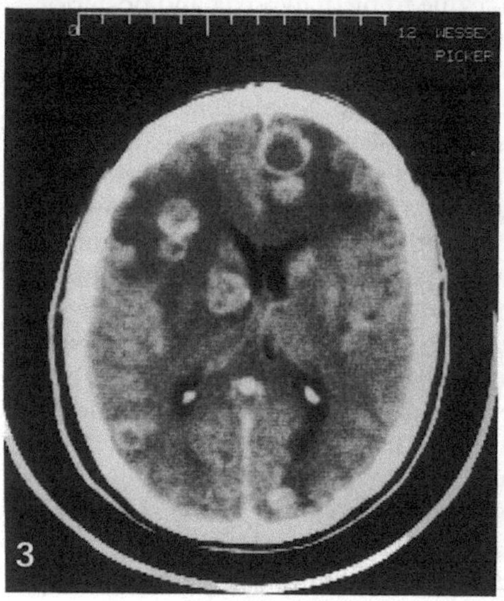

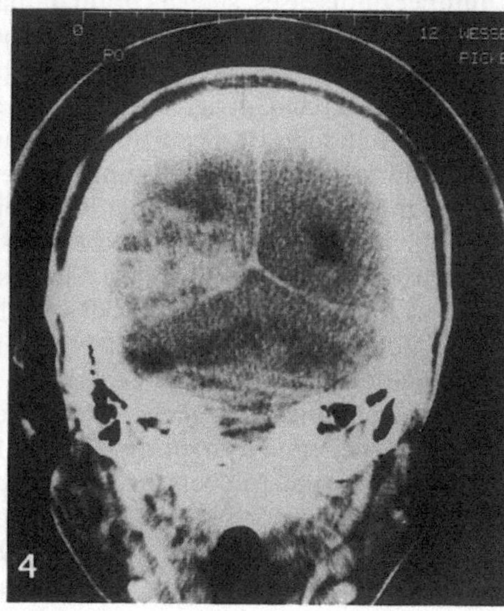

Fig. 1. CT scan showing an anaplastic astrocytoma as a dark area in the cerebral hemisphere posteriorly on the right. A calcified focus, medially, appears white

Fig. 2. Enhanced CT scan of the same patient showing a white enhancing rim to the tumour. Surrounding oedema (*black*)

Fig. 3. CT scan: multiple enhancing metastases (*white*) and surrounding oedema (*black*)

Fig. 4. CT scan: coronal reconstruction showing an enhancing meningioma arising from the upper surface of the tentorium cerebelli

of contrast medium to be injected into sensitive cerebral arteries. Angiography is still essential for the demonstration of arteriovenous malformations and for localising saccular aneurysms on major branches of the circle of Willis. Myelography is still widely used for the localisation of compressive lesions of the spinal cord and caudá equina.

The resolution of CT scanning of the brain and spinal cord continues to improve and this technique is particularly valuable for the demonstration of intracranial and intraorbital pathology especially as reconstructions of the brain and surrounding structures can be made in the horizontal, sagittal and coronal planes (Figs. 1–4). Ventricular dilatation and distortion of the cerebral hemispheres, brain stem, cerebellum or spinal cord are well demonstrated. Lesions, such as infarcts, abscesses and tumours, within the brain can often be localised by changes in density and thus changes in the attenuation of x-rays. In Fig. 1, an anaplastic astrocytoma in the right cerebral hemisphere is seen as a black area of decreased attenuation. A calcified area in the tumour is white. Following the intravenous injection of an iodine-containing medium such as Iohexol, the glioma enhances (Fig. 2) due to the leakage of contrast into the tissue from the abnormal tumour vessels. The tumour then appears as a cystic lesion with a white rim of enhancing tumour. The surrounding oedematous tissue is black. In the normal brain, there is a blood-brain barrier which prevents leakage of contrast medium into the tissue. Only areas such as choroid plexus and pineal, which possess no blood-brain barrier, are enhanced in the normal brain. Metastatic carcinomas (Fig. 3), Schwannomas and meningiomas (Fig. 4) are all tumours which are not derived from brain tissue, and these tumours all enhance on CT scan. Gliomas, on the other hand, usually only enhance if they are poorly differentiated; the degree of enhancement can, to some extent, be correlated with the histological structure of blood vessels within poorly differentiated gliomas. Other enhancing lesions include abscesses and early infarcts.

One major impact of CT scanning is that clinicians are able to determine the site and often the nature of the pathological lesions in the central nervous system with more certainty. Thus the effects of CT scanning upon the management of patients with cerebral tumours was initially to reduce the number of patients who were biopsied. However, neither CT scans nor MRI are reliable enough to ensure a firm histological diagnosis so that biopsy confirmation continues to be essential. In some centres, multiple CT scan guided biopsies are taken to allow more accurate assessment of the nature of different areas of the tumour.

4 Extracting Information from a Neurological Autopsy

There are two major factors involved in planning a neurological autopsy. First: the pathologist should be aware of the exact reasons for doing that particular autopsy. Is it a medico-legal autopsy to analyse the effect of a head injury and should photographs be taken for evidence? Or, is it a diagnostic autopsy to ensure that the causative organisms of a viral encephalitis, bacterial meningitis or abscess can be identified? Is it to give genetic counselling so that the autopsy should, as far as possible, be designed to determine whether the neurological disease is hereditary or sporadic? Is the autopsy important for the clinical diagnosis and correlation of clinico-pathological features? Will the case be published and thus require careful pictorial documentation of the pathology, not only of the brain, spinal cord, peripheral nerves and muscle, but also of other organs. Second: the nature of the pathological lesion and its site should be determined as far as possible by careful review of the clinical history. A few minutes spent in planning a post mortem examination may save many hours trying to salvage a diagnosis from an inadequately documented autopsy.

Techniques for the removal of the brain and spinal cord are described in detail by ADAMS and MURRAY (1982) and by LUDWIG (1979). Precautions to be taken against the spread of infection from viruses diseases and Creutzfeldt-Jakob disease are specified in a Code of Practice and Special Report published by the Department of Health and Social Security (1978 and 1981).

The fragility of neurological tissue should be recognised so that the brain, spinal cord and peripheral nerves are very carefully handled especially in the unfixed state in order to avoid unnecessary artefact. If there is a neuropathological problem, it is rarely solved by slicing the brain fresh. The brain and spinal cord should be suspended in 10% neutral buffered formalin for 3 to 4 weeks and examined in the hardened, fixed state so that anatomical relationships of the pathological lesions can be determined. Samples of tissue for virology, bacteriology or for biochemical estimation can be taken from the fresh brain before it is fixed.

Although much of the pathology may be investigated in the fixed brain, there are a number of important observations that can only be made at the time of the autopsy. Before the brain and its superior covering of dura are removed from the skull, the dural cupola should be reflected from the surface of the hemispheres so that any subdural haemorrhage, subarachnoid haemorrhage or flattening of the cerebral gyri due to brain swelling can be appreciated. As the brain is removed, the cranial nerves can be examined and, if necessary, traced into the dura or bony canals in the base of the skull. The dural venous sinuses, intracranial portions of the internal carotid arteries should be examined at the time of autopsy for the presence of thrombus. Detailed examination of the pituitary can be performed either by removing just the gland, or, more satisfactorily, by removing the whole sphenoid bone and dissecting the pituitary fossa after fixation.

In cases of optic nerve or orbital disease, the orbit and optic canals can be approached with ease from inside the skull. The middle ear can be examined through the tympanic part of the petrous temporal bone or by removal of the whole petrous temporal bone and examination of the fixed specimen. Assessment of skull fractures usually requires stripping of the dura; fractures are usually straight and haemorrhagic whereas sutures are irregular and jagged. Fractures of the cervical spine are best assessed in a fixed specimen; the whole cervical spine may be removed by disarticulating the atlanto-occipital joint at the base of the skull and cutting through the disc space just above T1.

Following its removal, the brain weight should be accurately recorded as a heavy brain is often an indication of cerebral oedema, tumour or a large intracerebral blood clot. Reduction in the brain weight often indicates cerebral atrophy due to Alzheimer's disease, old cerebral infarction, or a developmental abnormality. It should be repeated that the brain is very delicate and should be removed with great care. The midbrain and cerebral peduncles may be torn due to excessive traction on the cerebral hemispheres during removal of the brain, especially if this region is infarcted or involved by tumour. Such damage can be avoided if the posterior part of the skull incision is placed as far down the occipital bone as possible and all the structures attaching the brain stem to the walls of the posterior fossa are cut before removal of the brain is attempted. Traction on the midbrain is avoided by placing the fingers of one hand under the tonsils of the cerebellum and applying gentle traction while the other hand supports the cerebral hemispheres from underneath. The other area of brain which is usually damaged during removal is the lower medulla, mainly due to cutting the brain stem too high. Ideally the spinal cord should be cut through the second or third cervical segment at the same time that the vertebral arteries are detached from the walls of the spinal canal. Once the brain is removed it should be suspended in a bucket of formalin as soon as possible by a hook or a piece of string passed posterior to the basilar artery in the mid pontine region. Placing a fresh brain on a flat surface for any length of time will cause irreparable distortion.

The spinal cord and the vertebral column can be removed *in toto* and the cord subsequently extracted when the whole specimen has been fixed. This is an ideal method for examining spinal lesions and their effect on the cord. Alternatively the spinal cord alone may be removed by the anterior or posterior approach. The anterior approach by which the cord is exposed by sawing through the pedicles of the vertebrae, is probably easier and also allows spinal ganglia to be obtained at all levels. Care should be taken not to squash the cord by applying excessive traction. If the spinal dura is gripped at the level of the cauda equina, the cord can be progressively released by cutting along the spinal roots and removing the spinal ganglia. The cord, surrounded by its dura, can then be suspended in a long glass tube in buffered formalin with a small weight suspended from the lower end of the dura. Specimens of unfixed spinal cord for biochemical, virological and bacteriological investigations may be taken before the rest of the cord is fixed.

Histological examination of peripheral nerves and autonomic ganglia may be an essential part of the autopsy. Samples of the nerve, sympathetic chain and splanchnic nerves can be preserved by placing them on a piece of card so that they are slightly stretched; each end of the nerve can be lightly pressed onto the card so that it sticks when the tissue is placed in formalin. Specimens of muscle for paraffin embedding can be treated in the same way but frequently the assessment of muscle disease at autopsy is more satisfactory if samples of muscle are taken in the same way as a biopsy (see section 8) and the specimen frozen in liquid nitrogen for histochemical investigation.

5 Examination of the Fixed Brain and Spinal Cord

Although it may not be practicable in a busy general pathology service to fix every brain, it is, in most instances, advisable to fix the brains from patients with neurological problems if a reliable answer is to be obtained.

Just as the autopsy should be planned on the basis of the clinical data available, examination of the fixed brain also takes into account the clinical features and the results of post mortem examination of the other organs in the body. In particular, associated diseases which affect the brain such as hypertension, the presence of a mural thrombus in the heart, severe atherosclerosis in the carotid vessels, or the presence of a tumour elsewhere in the body should be noted.

5.1 External Features of the Brain

Examination of the brain itself should involve not only looking at the brain but also feeling it. Once the sagittal sinus has been examined for thrombus and the dura removed from the brain, softening of the surface, indicating recent infarction or the presence of a tumour, may be easily appreciated by carefully palpating the surface of the brain. Similarly, flattening of the gyri or gyral atrophy with sulcal widening may be detected by palpation. Discolouration of the surface of the brain by contusion, subarachnoid haemorrhage, meningitis or the presence of a tumour should also be sought.

In general, normal brains are symmetrical; if asymmetry is present it may be due to enlargement of one hemisphere or atrophy of the other. If there is enlargement of one hemisphere due to the presence of an intracerebral haematoma, tumour, abscess or oedema there may be shift and herniation of the brain. The different sites of herniation are determined partly by the position of the intracranial dural sheets which divide the cranial cavity into 3 major compartments and partly by the position of the expanding mass in the brain. The falx cerebri partly separates the two cerebral hemispheres in the midline and the tentorium cerebelli separates the cerebral hemispheres above from the cerebellum in the posterior fossa below. A space occupying lesion in one hemisphere may push the cingulate gyrus on the medial aspect of the cerebral hemispheres under the falx. More significantly, an expanding brain mass may cause the parahippocampal gyrus on the medial aspect of the temporal lobe to herniate into the opening in the tentorium cerebelli. Such transtentorial herniation may cause midbrain compression and fatal haemorrhage into the midbrain. The same herniation may also compress branches of the posterior cerebral artery against the free edge of the tentorium cerebelli as they pass across the inferior aspect of the temporal lobe; this results in infarction of the visual cortex and the inferior aspects of the temporal and occipital lobes. Expanding lesions in the cerebellum or elsewhere in the posterior fossa may result in prolapse of the cerebellar tonsils into the foramen magnum (coning) resulting in sudden death of the patient. Such herniation is usually recognisable because the tips of the cerebellar tonsils are soft, necrotic and pressed firmly against the back of the medulla.

Cranial nerves and arteries at the base of the brain are best examined before the brain is cut. Compression of the third nerve due to transtentorial herniation may be recognised by a transverse band of haemorrhage across the nerve; such third nerve compression may account for pupillary signs noted before the death of the patient. Inspection of major arteries arising in the circle of Willis may reveal thrombotic occlusion or saccular aneurysm, the latter most commonly arise from major branches of the anterior part of the circle of Willis.

5.2 Cutting a Brain

Instruments for cutting brains are standard equipment in most laboratories. As seen displayed in Fig. 5 on a non-slip cutting board approximately 0.5 × 0.75 m, the essential instruments consist of a long, thin sharp knife, scalpel, scissors, forceps, probe and ruler. The 1 cm – thick brass angles fit snugly around the brain for cutting 1 cm thick slices. A fish-slice is useful for picking up fragile brain slices and the pot of Indian ink is used for marking histological blocks on the reverse side to the cutting face.

Before the cerebral hemispheres are sliced, the midbrain should be cut cleanly and horizontally (Fig. 6); a 5 mm thick block of midbrain containing the red nuclei, substantia nigra and periaqueductal cranial nerve nuclei can then be removed and examined histologically if necessary. This area of the brain is often damaged by an unwary, curved cut; it is not possible to perform histology on a curved surface. Removal of the brain stem will allow any parahippocampal gyral herniation to be evaluated and then the cerebral hemispheres can be cut in a horizontal, coronal (Fig. 7) or sagittal plane depending upon the structures requiring demonstration.

Certainly, the anatomical arrangement of the internal capsule with its anterior and posterior limbs and genu are much better appreciated in a horizontal section, as is the anatomy of the lateral ventricles and the positions of the

Fig. 5. Brain cutting equipment (see text)

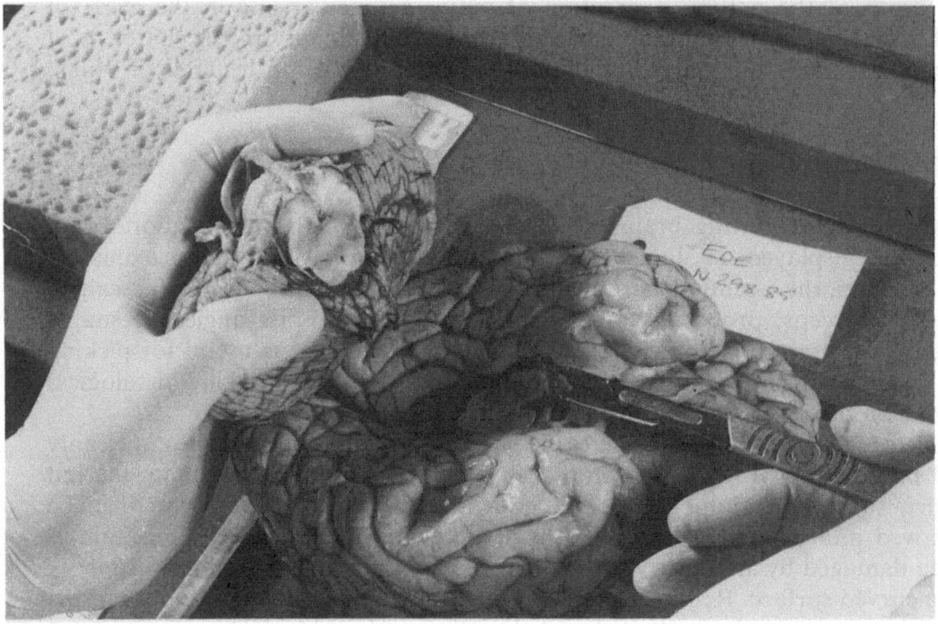

Fig. 6. Cutting through the midbrain with one sweep of the scalpel held like a pencil

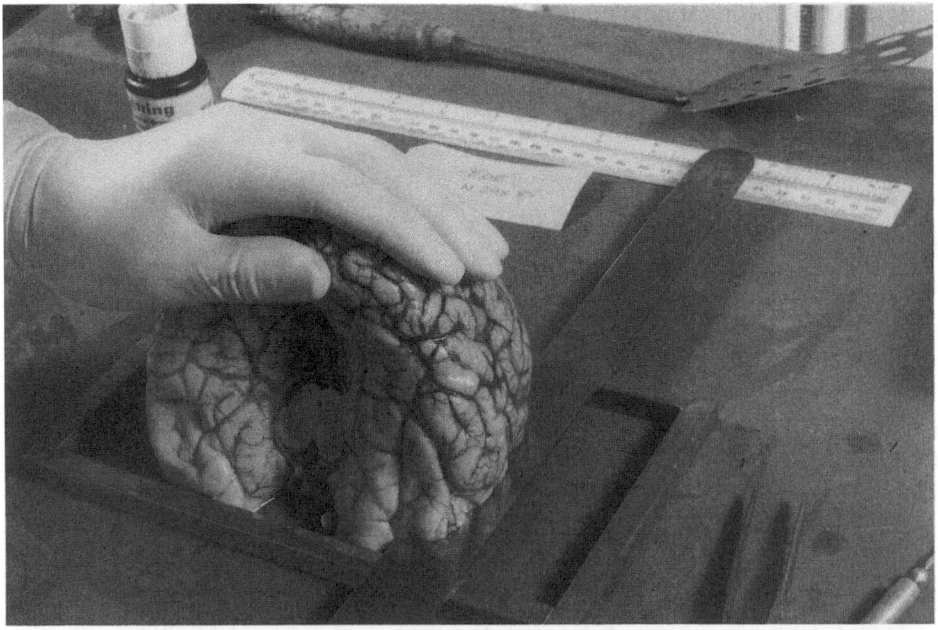

Fig. 7. Cutting the cerebral hemispheres using the 1 cm thick brass angles to obtain slices of even thickness

caudate nucleus, putamen, globus pallidus and thalamus with their relationship to the internal capsule. Brain shift and herniation with distortion of the ventricles, on the other hand, are much better appreciated in coronal slices as is the relationship of the internal capsule of the cerebral peduncles and the anatomy of the hippocampus and lateral geniculate bodies. It is also easier to define the arterial territories of the anterior, middle and posterior cerebral arteries in coronal sections and to appreciate the source of intracerebral haemorrhage. Most massive intracerebral haemorrhages in hypertensive patients occur in the region of the basal ganglia and the internal capsule. Intracerebral haemorrhages arising from ruptured saccular aneurysms, however, may cause some confusion but they can usually be traced either to the region of an anterior communicating artery aneurysm or to a middle cerebral aneurysm in the lateral fissure. The appearance of blood in the subarachnoid space may similarly cause confusion. It may be primary, due to a ruptured saccular aneurysm, or to bleeding from a vascular malformation on the surface of the brain or due to trauma and contusion. Alternatively, bleeding in the subarachnoid space may be secondary, and result from blood leaking through the foramina of Luschka following an intraventricular haemorrhage.

Sagittal sections of brain are valuable for the examination of midline lesions such as pituitary tumours or tumours involving the fourth ventricle.

There are two standard ways of sectioning the brain stem and cerebellum. Horizontal cuts can be made through the pons and cerebellum, or medulla and cerebellum to assess the size of the fourth ventricle or its contents. Alterna-

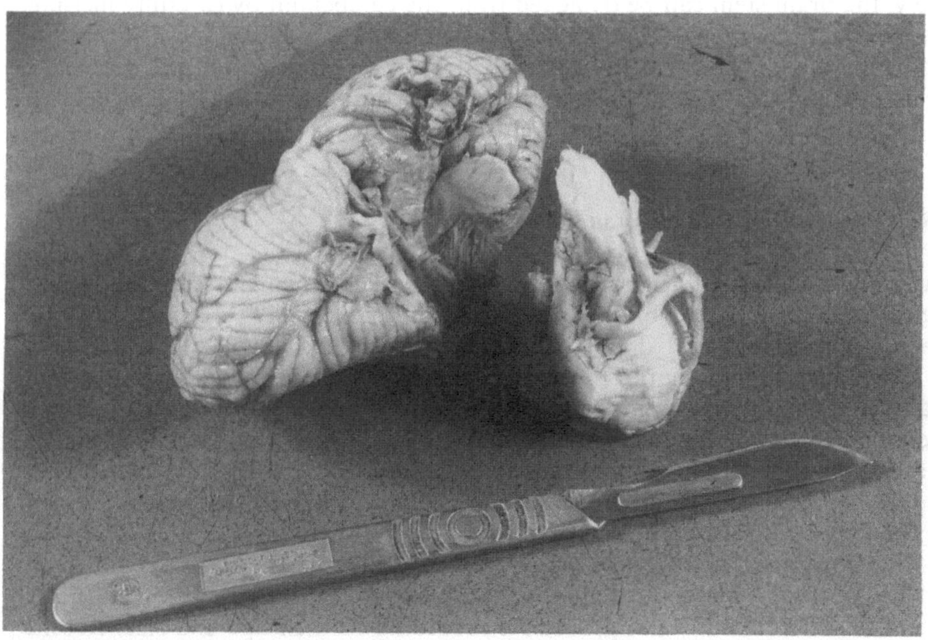

Fig. 8. The brain stem has been detached from the cerebellum by cutting through the peduncles with a scalpel

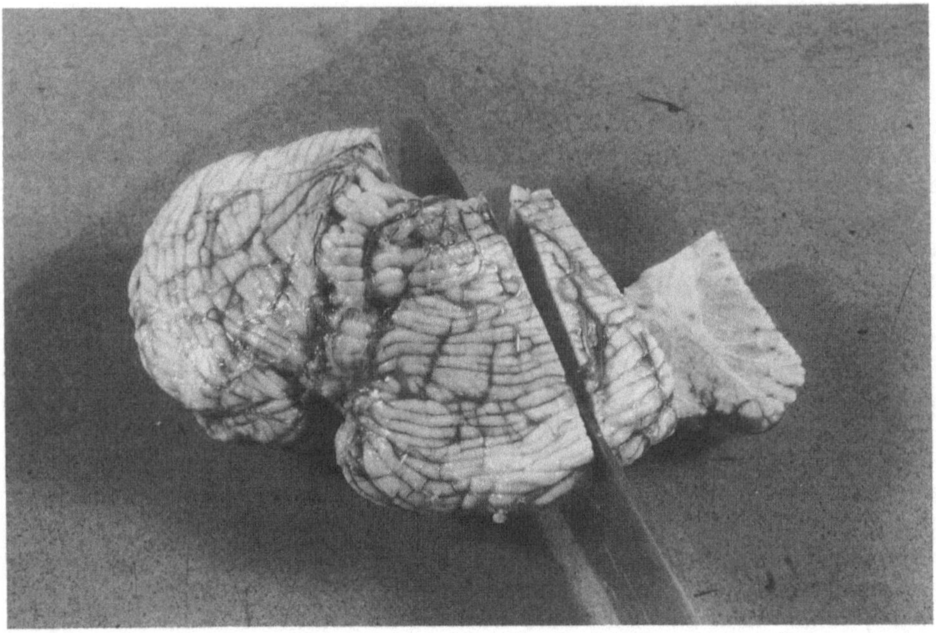

Fig. 9. Radial sections of the cerebellum are cut at right angles to the folia

tively, the brain stem can be detached from the cerebellum by carefully incising the three cerebellar peduncles on each side (Fig. 8). Parasagittal slices of cerebellum (Fig. 9) are ideal for demonstrating atrophy or infarction of the cerebellar folia. The brain stem can be cut in 0.5 cm horizontal slices to reveal structures in the pons and floor of the fourth ventricle.

5.3 Assessment of Pathology in a Cut Brain

There are a number of general questions to be asked when assessing the pathological changes in a cut brain. Is the brain asymmetrical? Are there obvious macroscopic lesions and if not, from which areas should histological blocks be taken in order to characterise the pathology? It is unusual for the whole brain to be affected symmetrically by a disease process except when there has been total cerebral hypoxia as in a patient kept alive on a ventilator with inadequate cerebral perfusion. In these cases the brain is universally soft, discoloured and autolytic. Areas of normal brain are present in most other cases and can be compared with the pathological lesions.

Haemorrhages will be obvious as fresh red blood clot or granular brown old haemorrhage but the extent of cerebral infarction may be more difficult to define. Infarcts differ in appearance according to their age and size. Recent infarcts, 24–72 hours old, are characterised by petechial haemorrhages in the cortex and pale, soft underlying white matter. In very large infarcts, the haemorrhagic areas may be confined to the edges of the infarct as the petechial haemorrhages are thought to be due to leakage of red cells following reflow of blood from neighbouring territories through infarcted, damaged vessels. Infarction may be confined to individual arterial territories in the brain or may involve the boundary zones where adjacent arterial territories meet. Boundary zone infarcts (BRIERLEY and GRAHAM 1984) are commonly caused by periods of hypotension in older people. Within 2 or 3 days, infarcted brain tissue becomes soft and oedematous and starts to disintegrate. Removal of tissue from cerebral infarcts is slow, particularly if the damaged area is large. The classical fluid-filled cystic infarct may not be formed for a number of months or years.

Small recent infarcts caused by minute emboli arising from a mural thrombus in the heart, or from atherosclerotic plaques in the neck vessels, may be seen as slightly shrunken soft patches in the white matter, or small haemorrhagic areas in the cortex, particularly in the depths of sulci. Although some estimate of the age of an infarct can be made from its macroscopic appearance, a more accurate assessment is usually obtained from the clinical history or from histology (see Sect. 6.2.3).

Tumours in the brain may be identified by the abnormal consistency and colour of the tissue, or by the mass effect of the tumour and the surrounding oedema causing distortion of the brain anatomy. Poorly differentiated gliomas such as glioblastoma multiforme often contain areas of necrosis, have irregular profiles and appear to be well-circumscribed. The borders of well-differentiated astrocytomas in the cerebral hemispheres, pons or cerebellum, however, may be difficult to define as the colour of the tissue and sometimes its consistency is similar to that of normal brain tissue. Metastatic tumours are usually multiple, well-circumscribed, and often circular in profile and spherical in shape.

Other macroscopic lesions may be very obvious. An abscess, for example, has a central mass of pus surrounded by a rim of soft haemorrhagic tissue in the early stages or a fibrous capsule in the more chronic lesion. The firm grey gliotic plaques of longstanding multiple sclerosis, are frequently distributed around the lateral ventricles, cerebral aqueduct and throughout the brain stem (MATTHEWS et al. 1984). Detection of smaller plaques at the junction of cerebral cortex and white matter may require a careful search. Recent multiple sclerotic plaques are soft and granular, and although they may be confused macroscopically with infarcts, they can easily be distinguished histologically by the preservation of axons within the plaque.

Contusions, haemorrhages and shearing lesions in the white matter of brains following head injury may be visible macroscopically but, the major brain damage in a head injury is often diffuse axonal injury and this may only be detectable by histology (ADAMS, this volume). Similarly, the brains of patients with encephalitis may show few macroscopic changes except for cerebral oedema. In some cases, as in herpes simplex encephalitis, the pathology is concentrated in certain parts of the brain and histological evaluation will be directed to those areas (ANDERSON, this volume).

Cerebral oedema due to breakdown of the blood-brain barrier and the outpouring of fluid is commonly seen around tumours and infarcts and may produce focal or generalised brain swelling. The oedema fluid accumulates mainly in the white matter producing softening, swelling, and widening of cerebral gyri.

Ventricular dilatation in the cut brain requires careful evaluation. The cerebrospinal fluid pathways should be followed from the lateral ventricles through the third ventricle and aqueduct to the foramina of Luschka and the subarachnoid space in the search for an obstruction. In obstructive hydrocephalus all the chambers proximal to the cerebrospinal fluid block will be dilated. If ventricular dilatation is due to atrophy of brain tissue, on the other hand, no obstruction will be detected and the cerebral aqueduct and fourth ventricle will be normal in size.

In some diseases, there may be little asymmetrical change in the brain. Cerebral atrophy in Alzheimer's disease, for example, may be symmetrical and involve much of the frontal lobes on both sides. In Parkinson's disease, the loss of pigmentation of the substantia nigra may be symmetrical, and in Huntingdon's chorea, atrophy of the caudate nuclei putamen and globus pallidus may also be symmetrical. In these conditions, it is often wise to compare the appearances with a normal brain.

Examination of brains from very young children may present some difficulty mainly because they are soft and friable, but also because the lack of myelin within the cerebral hemispheres makes anatomical landmarks difficult to detect. Subependymal haemorrhages in premature infants are usually obvious and areas of cortical and white matter infarction, as a result of birth hypoxia, appear as softened, granular or cystic lesions. If children survive for many months following birth hypoxia, there may be extensive cyst information in the cortex and white matter (FRIEDE 1975). Gross disorganisation of the brain usually suggests damage in early fetal life but abnormalities of gyrus formation, e.g. polymicrogyria, suggest damage during later fetal life. The distinction between genetic disorders of the brain and sporadic disorders usually depends upon correlation of the clinical features, cytogenetic information, family history and pattern of brain damage.

5.4 Spinal Cord

Diseases of the spinal cord may affect virtually the whole length of the cord as in motor neuron disease or, there may be a single focal lesion such as an intrinsic tumour of the cord or compression by an extrinsic tumour. There may be several focal lesions, as in multiple sclerosis or multiple metastatic deposits. The clinical history and examination is usually essential for establishing the level of the cord involved. It must be remembered, however, that the lower end of the spinal cord is at the level of L1 vertebra so that the segmental levels in the cord differ from those in the vertebral column.

Once the dura of an isolated spinal cord has been opened, the anterior and posterior surface of the cord can be identified by the presence of a single anterior spinal artery and by an irregular venous plexus on the posterior aspect of the cord. The cervical enlargement marks the upper end of the cord, and the lumbar enlargement and cauda equina are at the lower end. Segmental levels can be calculated by first identifying the posterior roots of the first thoracic segment, which are usually the last large roots arising from the cervical enlargement; T2 roots below it are much smaller. Segments above and below T1 can be calculated by identifying and counting the posterior roots at each level. If no focal lesion of the cord can be identified macroscopically, the site of the lesion can usually be traced by the pattern of long tract degeneration in the cord in histological sections. Above the lesion, there will be degeneration and atrophy of the dorsal columns whereas below the lesion the cortico-spinal tracts will be damaged (see Sect. 6.1). Such degeneration can be detected by the loss of myelin staining or by the location of myelin breakdown products within the damaged tracts.

5.5 Artefact

Failure to recognise artefact in post mortem brain and spinal cord may lead to unnecessary histological investigation and even misdiagnosis. Damage may occur to the surfaces of the cerebral hemispheres, the midbrain may be torn, or the corpus callosum may be severed by the free edge of the falx cerebri during removal of the brain from the skull. Post mortem bacterial growth within inadequately fixed or even in well-fixed brain, may produce focal red or black discolouration; occasionally gas bubbles form within the white matter deep in the hemispheres. It is wise to check the formaldehyde concentrations if these features do occur. Microscopically, post mortem bacterial colonies appear as basophilic granular clumps with no accompanying inflammatory reaction.

The unfixed spinal cord is very soft and fragile and particularly prone to artefact from harsh handling. It may be squashed and flattened during removal or intussusception may occur with one segment of cord squeezed into an adjacent segment. If there is doubt about the significance of the macroscopic appearance it can always be checked by histology.

5.6 Selection and Processing of Tissue Blocks for Histology

Selection of blocks of brain for histology should take note of the clinical features, and thus the localisation of the pathological process, the macroscopic appearances of the brain, and a desire to retain the anatomical identity of the blocks. If, for example, a piece of frontal lobe is taken (Fig. 10), it is advisable to include a small part of corpus callosum for orientation. A block of hippocampus should be taken at a level which adequately shows the typical shape of Ammon's horn, and blocks of brain stem and cerebellum should allow the anatomical location to be identified once the sections have been processed and stained. In those brains in which the pathological lesion is macroscopically identifiable, it is often useful to take blocks which include normal as well as pathological tissue for comparison. When there is no obvious macroscopic lesion, the choice of blocks should be determined by the clinical features of the disease. If no localising neurological signs were present clinically, the selection of blocks

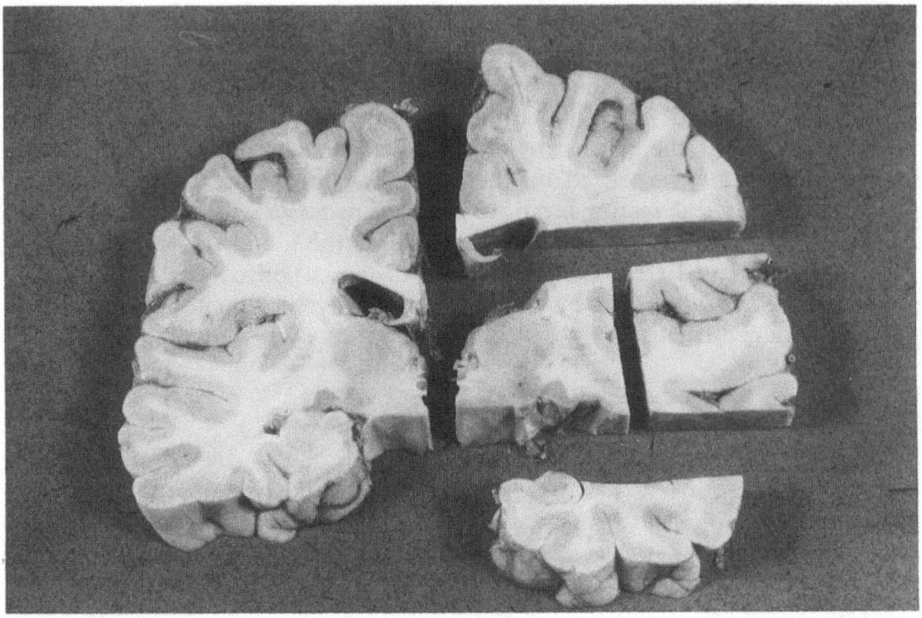

Fig. 10. Blocks for histology: anatomically recognisable blocks of frontal lobe, thalamus and hippocampus

should be guided by previous accounts of similar diseases. For example, certain areas of the brain such as the hippocampus and cerebellum are very sensitive to hypoxia so it is from these regions that blocks are taken to confirm hypoxic damage to the brain.

Blocks of brain can be dehydrated and embedded either in paraffin wax or in celloidin. Paraffin wax has the advantage that it is a shorter embedding schedule, although with special vacuum techniques, the length of celloidin embedding schedules can be reduced. Blocks of the size shown in Fig. 10 may take twelve days or more to dehydrate prior to embedding but 5–20 μm paraffin sections of such blocks can be cut on a standard sledge microtome. Large blocks including both cerebral hemispheres (Figs. 11, 12) may require a larger microtome. Small blocks such as transverse sections of spinal cord may be processed on a routine 24 hour cycle. The main disadvantage of paraffin embedding is the 30% or so shrinkage of the tissue during dehydration. Very much less shrinkage occurs with celloidin embedding so this technique may be more suitable for quantitative work. Nevertheless, celloidin sections are thicker; they are cut at 15–20 μm so that although they are ideal for low power, high resolution microscopy is difficult. Celloidin sections are particularly useful for detecting disruption of neuronal patterns within the cerebral cortex or elsewhere in the brain. However, there are some techniques which cannot be applied to celloidin sections so paraffin sections usually allow a greater flexibility of histological techniques.

6 Histological Techniques in the Recognition of Pathology in the Central Nervous System

Although paraffin sections of brain and spinal cord stained by general histological techniques (Table 1) are very useful for detecting primary or metastatic tumours, and for defining areas of inflammation and blood vessel proliferation,

Table 1. General histological stains in neuropathology

Haematoxylin and eosin

Stains nuclei blue, cytoplasm rich in RNA blue (e.g. Nissl granules and regenerating muscle fibres): cell processes (neuropil) and connective tissue pink; myelin pink-blue; calcification blue.

Main uses

Demonstration of the form and number of nuclei within the tissue. Neurons may be distinguished by vesicular nucleus, prominent nucleolus and basophilic (RNA rich) cytoplasm. Hyperchromasia and pleomorphism in tumour cells is well shown. Inflammatory cell accumulations are well seen.

Limitations

Cell processes, axons, myelin and connective tissue in the brain are not well demonstrated.

Haematoxylin van Giesen

Stains connective tissue collagen red; nuclei black; myelin black; cytoplasmic processes dull yellow.

Main uses

Detection of collagen-free brain tissue within connective tissue structures. Valuable stain for nuclear form in CNS tumours; detection of myelinated fibres; detection of thick collagen bands in meningiomas; collagen in abscess walls.

Limitations

Small collagen fibres are not well shown and neural cell processes are not distinguishable.

Reticulin (Gordon and Sweet)

Stains small collagen fibres black.

Main uses

Combined with a neutral red counterstain the reticulin technique is ideal for detecting connective tissue in abscess walls, around blood vessels in normal brain and in tumours; demonstrates capillary endothelial proliferation in poorly differentiated gliomas. Connective tissue surrounding cells in Schwannomas is useful for distinguishing these tumours from most types of meningiomas. Detection of connective tissue in gliosarcomas.

Limitations

Nuclei and cytoplasmic features not well demonstrated.

Periodic acid Schiff

Main uses

Stains glycogen and mucins pink.
Identification of glycogen in muscle fibres and metastatic tumour cells. With appropriate diastase controls, detection of mucin in metastatic tumour cells. Combined with Orange G, detects the admixture of cells which distinguishes normal pituitary from pituitary adenoma.

Table 2. Histological demonstration of cellular and tissue elements in the normal and pathological nervous system

	Major uses	Comments
Neurons		
Nissl (Cresyl violet)	Stains Nissl granules (RNA and endoplasmic reticulum) in neuronal cytoplasm. Detection of chromatolysis. Detection of neuronal groups at low power	Small neurons may be difficult to distinguish from glia. Dendrites and axons are not stained
Silver stains (Palmgren, Bodian, Glees and Marsland)	Stains axons, axon balloons, neurofibrillary tangles and neuritic (Alzheimer) plaques	Can be capricious and expensive
Neurofilament proteins	Detection of axons in the central and peripheral nervous system	Immunoperoxidase technique: does not stain neuronal cell bodies
Neuron specific ($\gamma\,\gamma$) enolase	Stains neuronal cytoplasm	Not specific for neurons in pathological material
von Braumühl	A reliable technique for the detection of neuro-fibrillary tangles and plaques	Frozen sections
Golgi technique	Stains individual neurons and processes. Ideal for studying dendritic arborizations	Requires staining of neurons in the block and thus limits the use of the tissue for other techniques
Myelin		
Luxol fast blue/ Nissl (Klüver Barrera technique)	Stains myelin blue; combined with the Nissl technique also stains neurons. Very useful technique for detecting oedema, demyelination and tract degeneration	Does not stain axons *but*, combined with the Palmgren stain axons and myelin can be demonstrated in the same section
Loyez	Stains myelin black	
Oil-Red-O	Stains the hydrophobic lipids (mainly cholesterol esters) in degenerating myelin	Frozen sections are required and myelin breakdown products are only in the tissue for a limited period after the damage
Marchi technique	Degenerating myelin is stained black. Suitable for tracing tract degeneration	A straightforward technique but requires staining in the block and thus serial sections are not available for other techniques
Oligodendrocytes		
Carbonic anhydrase C	Stains the cytoplasm of mature oligodendrocytes in paraffin sections	Immunocytochemical technique. May require special fixation of tissue in Carnoys fixative; is not a reliable marker for oligodendrogliomas
Penfield	Detects oligodendrocyte processes	Silver technique and capricious

Table 2 (continued)

	Major uses	Comments
Astrocytes		
Glial fibrillary acidic protein	Stains fibrous astrocytes in normal white matter; reactive astrocytes in damaged brain tissue; and cells in glial tumours	Immunoperoxidase technique. Best stain for fibrous and reactive astrocytes. Does not demonstrate protoplasmic astrocytes in grey matter well
Holzer technique	Stains fibrous and reactive astrocytes purple	A capricious stain. Good for low power microscopy
Phosphotungstic acid haematoxylin	Stains reactive astrocytes and tumour astrocytes blue	Also stains myelin blue and thus only suitable for grey matter. A capricious stain
Cajal gold sublimate technique	Stains astrocytes in the white and grey matter black. Stains reactive astrocytes	Frozen sections required
Choroid Plexus		
Carbonic anhydrase C	Stains normal choroid plexus and papilloma epithelium but not ependyma or metastatic carcinoma cells	Immunoperoxidase technique
Blood Vessels		
Factor VIII related antigen	Identification of blood vessel endothelial cells	Immunoperoxidase technique; usually unsuitable for post mortem material
Reticulin stain (Gordon and Sweet)	Stains the fine collagen surrounding blood vessels	Other tissue elements are not usually well demonstrated
Microglia and Macrophages		
Alpha-1-antichymotrypsin	Stains the processes of microglia and cytoplasm of foamy macrophages. Useful for tissue sections and for cytospin preparations of isolated cells	Immunoperoxidase technique: Does not work very well on post mortem material
Monoclonal antibodies for monocyte/macrophages	Reliable techniques for detecting microglia, macrophages and monocytes in the tissue	Immunoperoxidase techniques on cryostat sections
Hortega silver carbonate	Microglia	Capricious
Lymphocytes		
Monoclonal antibodies	Identification of lymphocyte subsets in inflammatory exudates and in lymphomas of CNS	Immunoperoxidase techniques mostly on cryostat sections
Schwann Cells		
S100 protein	Detects Schwann cells in peripheral nerves and in Schwann cell tumours	Immunoperoxidase technique not specific for Schwann cells; stains glia and neural crest derivatives

these techniques are not particularly suitable for unravelling the complexities of cellular inter-relationships in the central nervous system. It is difficult to detect small myelinated fibres and it is not possible to differentiate between dendrites, axons and astrocyte processes in the neuropil of the cerebral cortex and other areas of grey matter. For the last 100 years, considerable effort has been devoted to devising histological techniques in order to clarify cellular relationships in the normal and pathological brain (BANCROFT and STEVENS 1982) (Table 2). In planning a histological investigation, the advantages gained from the use of these techniques (Tables 1, 2) must be balanced against the complexity and capricious nature of some of the techniques.

In this part of the chapter, the normal histology and general pathological reactions within the central nervous system are briefly reviewed in order to illustrate the use of various techniques for detecting pathological damage to the nervous system. The development of immunocytochemical techniques during the last decade has solved many of the problems in identifying of cells and their processes in histological sections of brain and spinal cord but, new problems have arisen and these will be discussed later in a separate section on immunocytochemistry (see Sect. 6.4).

6.1 Low Power Microscopy of Large Histological Sections

In the description of gross brain pathology, it was stressed that blocks of brain should be taken for histology in such a way that the anatomy is still recognisable. This contact with the anatomical features should also be maintained while histological features are evaluated. As an extension of "holding the slide up to the light", low power microscopy under a dissecting microscope or hand lens is a very valuable, if not an essential, exercise in locating areas of damage in large sections of brain.

The most useful stains for low power microscopy are the Klüver Barrera technique for myelin (Fig. 11), the Holzer technique (Fig. 12) for astrocytes and a Haematoxylin and eosin stain for detecting accumulations of nuclei in tumours and in inflammatory lesions. Because the Klüver Barrera technique stains myelin dark blue and the grey matter pale blue, any distortion of the anatomy of the region, as, for example, by atrophy, or infiltrating astrocytoma, is well depicted. Furthermore, diffuse pallor of myelin staining may define areas of cerebral oedema, whereas sharply demarcated loss of myelin staining is seen in demyelinated multiple sclerosis plaques (Fig. 11). Axon and myelin loss from degenerated long tracts in the spinal cord may be demonstrated particularly well with the Klüver Barrera stain at lower power (Fig. 13). Other techniques for myelin such as Loyez's, serve a similar purpose.

As astrocytes form the scar tissue in virtually all pathological lesions in the brain, scarred areas may be detected at low power with the Holzer stain (Fig. 12). However, the Holzer stain is capricious and good results may be difficult to attain. Phosphotungstic Acid Haematoxylin (PTAH) stains reactive astrocytes and areas of gliosis blue, but as this technique also stains myelin

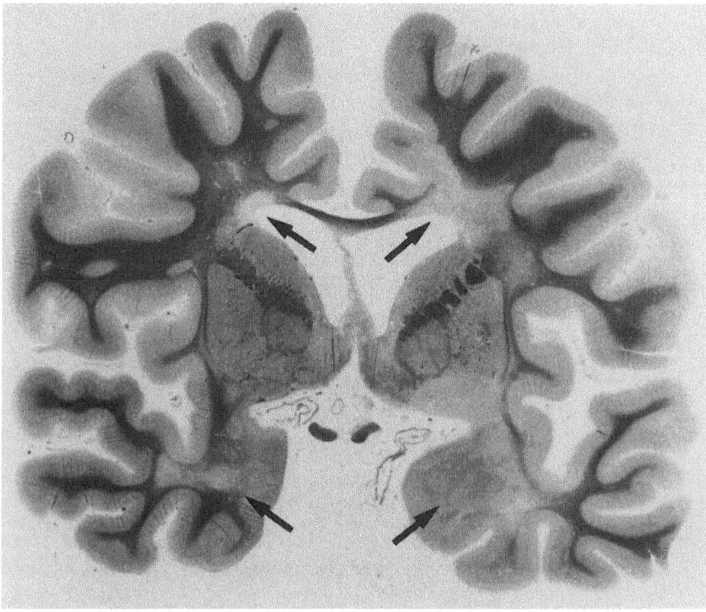

Fig. 11. Coronal section of both cerebral hemispheres. Demyelinated multiple sclerosis plaques (*arrows*). Klüver Barrera stain, ×0.85

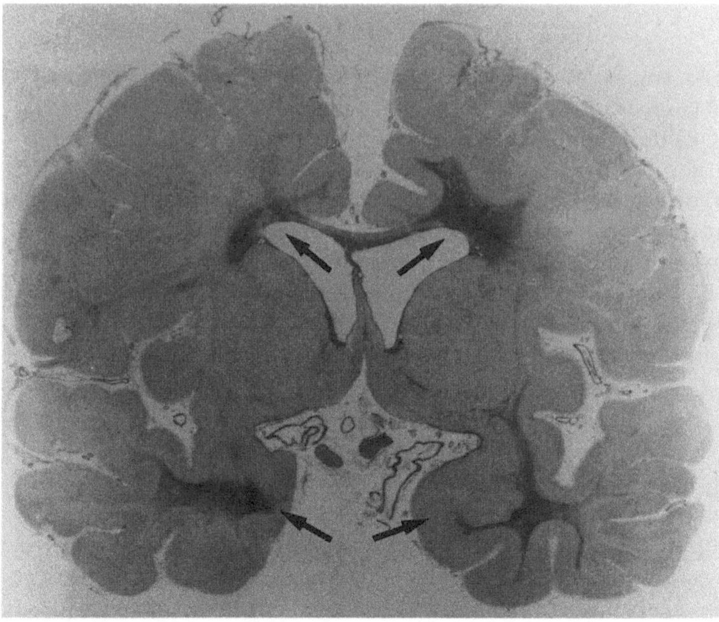

Fig. 12. Serial section of Fig. 11. Gliotic multiple sclerosis plaques (*arrows*) are heavily stained by the Holzer technique for astrocytes, ×0.85

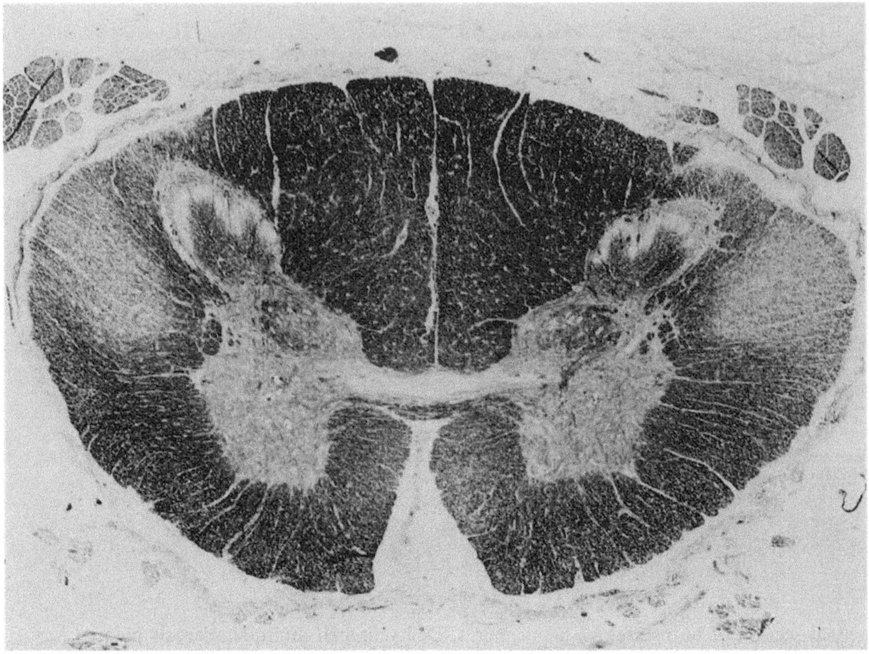

Fig. 13. Spinal cord in motor neuron disease (amyotrophic lateral sclerosis) showing loss of staining in the lateral cortico-spinal tracts due to loss of axons and myelin. Klüver Barrera, × 12

a similar dark blue, its use is limited to grey matter areas where it is ideal for detecting small areas of old infarction especially in cerebellar folia.

Diffuse loss of neurons may be difficult to detect in the central nervous system, especially if there is little astrocytic reaction. However, examination of 20 μm paraffin or celloidin sections stained with the Nissl (cresyl violet) technique allows disturbance of the normal regular neuronal arrangement in the cortex, hippocampus, lateral geniculate bodies and other areas to be detected. This technique is also useful for demonstrating the abnormal cortical histology in polymicrogyria as well as in areas of focal neuronal loss. Another technique that is useful for the detection of brain pathology at low power is Haematoxylin van Gieson. The normal brain contains a little collagen, but it is usually confined to blood vessel walls and to the meninges. Thus, collagenous tissue around abscesses, in some glial tumours and associated with arterio-venous malformations is well demonstrated by this technique. Haematoxylin van Gieson is also particularly valuable for the detection of neural tissue in teratomas, and in ectopic sites such as the dermis or subcutaneous tissue covering myelomeningoceles. In these sites, the neural tissue is a dull yellow and contrasts well with the surrounding red-stained collagenous connective tissue. Myelin and nuclei are stained black by the Haematoxylin van Gieson technique so this method serves as a very good general histological stain for the brain.

6.2 Medium and High Power Microscopy

Although much information can be obtained by viewing large sections at low power, medium and high power light microscopy is usually necessary both to analyse the changes that have been localised at low power and to detect those changes which were not immediately apparent at low power.

The nuclei of neurons, astrocytes, oligodendrocytes and ependymal cells can be detected in haematoxylin and eosin stained sections of normal brain, but demonstration of the processes of these cells requires other techniques (Table 2) (WELLER 1984a). Large areas of eosinophilic neuropil between nuclei and cell bodies in the cortex contain many synapses, dendrites and astrocyte processes which can be very well depicted by electron microscopy. With the light microscope, cytoplasmic astrocyte processes in normal cortex can be stained by the Cajal technique whereas fibrous astrocyte processes in the white matter are well demonstrated by the immunoperoxidase technique for glial fibrillary acidic protein. Oligodendrocytes in the white matter and cortex contain carbonic anhydrase C which can be identified by immunocytochemistry; oligodendrocyte processes can be visualised by the Penfield method. Nissl granules (RNA and endoplasmic reticulum) in neuronal cell bodies are basophilic in Haematoxylin and eosin stained sections, and blue/purple with the Nissl technique. Axons can be demonstrated by silver techniques or by staining for neurofilament proteins by immunocytochemistry (Table 2). There are many stains for myelin sheaths in paraffin sections and perhaps the most useful is the Luxol fast blue technique especially when combined with the Nissl method to produce the Klüver Barrera technique.

6.2.1 Pathology of Neurons and Axons

Neurons are very sensitive to hypoxia, particularly those neurons in the hippocampus, cerebellum and frontal cortex. The earliest changes that can be detected in a haematoxylin and eosin stained section following hypoxic damage, is the disappearance of basophilic Nissl substance from the cytoplasm leaving a neuron with brightly eosinophilic cytoplasm and a shrunken pyknotic nucleus. This change may be detected within 6–12 hours of hypoxic damage and is followed by disintegration of the neuron and phagocytosis of the debris by microglia (neuronophagia) and macrophages.

Damage to axons produces changes not only in the axon itself but also in the neuronal cell body. Following the severance of an axon by infarction, trauma or by invading tumour, the distal portion of the axon degenerates. Some 2–3 days after injury, the distal end of the proximal axon and the proximal end of the distal axon have swollen to form axon balloons some 50–100 μm in diameter. Such structures can be seen in haematoxylin and eosin stained sections as eosinophilic circular or oval profiles but are more readily detected by silver techniques such as Palmgren, Bodian or Holmes.

During the stage of axon balloon formation and usually beginning at around one week after axonal damage, chromatolysis is seen in the neuron cell bodies from which the axons arise. In many instances this is a regenerative phenomenon characterised by a rise in rate of protein synthesis and, despite the disappearance of Nissl granules, the amount of RNA in the cells is increased. But in other circumstances, swelling of the neuron in this way may be a degenerative reaction. Chromatolysis can be demonstrated either in haematolylin and eosin stained sections or more clearly in the Nissl or Klüver Barrera preparations. The cell body becomes swollen and pale and the nucleus eccentric. Despite the early regenerative features in both axons and neurons, axonal regeneration is largely ineffective in the central nervous system and after 2 or 3 weeks, the changes of chromatolysis start to subside and axon balloons shrink. Evidence of axonal degeneration is detectable 2 or more weeks after the damage through the presence of myelin breakdown products either by the Marchi technique or by Oil Red 0 staining. Eventually, the myelin breakdown products are removed by macrophages and after several months, they may be undetectable. At this late stage, therefore, localisation of axonal degeneration depends largely upon detecting loss of myelin staining in Luxol fast blue or Klüver Barrera stained sections (Fig. 13) and the loss of axons from silver stained sections.

The changes associated with Alzheimer's disease may be detectable in haematoxylin and eosin stained sections but are much more satisfactorily demonstrated by silver techniques (see SCHOLZ, this volume). Neurofibrillary tangles in neurons and the complex interweaving processes in neuritic plaques are well demonstrated by these techniques. Congo Red and Thioflavin S may also be used to demonstrate neurofibrillary tangles and are ideal for localising amyloid at the centre of neuritic plaques.

6.2.2 *Pathology of Myelin*

Demyelination must be distinguished from axonal degeneration as the pathological significance is very different (MATTHEWS et al. 1984). Demyelination occurs in multiple sclerosis, post infectious encephalomyelitis, certain virus infections such as progressive multifocal leucoencephalopathy and in a number of hereditary disorders such as metachromatic leucodystrophy and Krabbe's leucodystrophy. Myelin sheaths are selectively destroyed and the axons are relatively spared. Not all the axons remain intact, however, and there may be considerable axonal degeneration in demyelinated areas. Demyelinating lesions can be identified by staining serial sections, one with a myelin stain and the other with an axon stain or the two techniques may be combined in one section as a Palmgren-Luxol fast blue stain (WELLER 1984a).

It is not usually difficult to distinguish between a cystic area of infarction and a solid, gliotic demyelinated multiple sclerosis plaque in the cerebral hemispheres but in the spinal cord, the nature of the lesion may be more difficult to define. In transverse sections of spinal cord, demyelinated multiple sclerosis plaques appear as circular or triangular areas of myelin loss which are not confined to the anatomical boundaries of ascending or descending tracts and contain intact axons. Axonal degeneration, on the other hand, is typically located through the loss of myelin staining confined to anatomical pathways such as the dorsal columns or cortico-spinal tracts (Fig. 13).

6.2.3 Determining the Chronology of a Brain Lesion

Correlation of the pathology of a brain lesion with the clinical signs and symptoms depends largely upon the identification of histological damage at the appropriate site and of the appropriate age. This requires a firm appreciation of neuroanatomy and of the sequence of events that follows tissue damage. The principles involved are similar in many respects to those in other organs. If tumours are excluded, the majority of pathological lesions in the brain involve cell death, often with inflammation and scarring. But, as distinct from other organs in which scars are composed largely of collagen, scar tissue in the brain consists almost entirely of astrocytes and their processes. Fibrous capsules around cerebral abscesses are notable exceptions. The sequence of events that follows brain infarction can be regarded as a general model for dating lesions in the central nervous system. Some variation in the timing will be seen in as much that inflammation, cyst formation and gliosis proceed more slowly in the centre of large infarcts than in small infarcts. It is often advisable to examine the edge of a large lesion to gain the best indication of its age.

The initial changes in cerebral infarction can be recognised in the cortex by the presence of small perivascular haemorrhages and, after approximately 6–12, hours by hypoxic changes in the neurons. These features are well-demonstrated in haematoxylin and eosin stained sections with the hypoxic neurons showing eosinophilia of the cytoplasm and pyknosis of the nuclei. Also at this time other cells will show nuclear pyknosis and disruption of the cytoplasm. By 12 hours, tissue oedema is often detectable with swelling and pallor of the cortex and white matter well-demonstrated by the Klüver Barrera technique but also detectable in haematoxylin and eosin preparations. The oedema increases over the next 24 hours and spreads preferentially through the white matter.

As the tissue disintegrates, a few polymorphonuclear leucocytes will accompany the oedema fluid into the infarcted tissue but these cells are seen mainly at the edges of an infarct. At a similar time, microglia, the resident histiocytes in the brain, proliferate at the edge of the infarct; they can be identified by their short, blunt, rod-shaped nuclei in haematoxylin and eosin stained sections, and by staining their processes by immunocytochemical techniques for alpha-1-antichymotrypsin or monoclonal antibodies for monocyte markers. Within 3–7 days, the microglia at the edges of infarcts are supplemented by monocytes derived from the blood. As they ingest lipid-rich degenerating brain tissue, the monocytes and microglia become swollen to form foamy macrophages identifiable in haematoxylin and eosin stained sections by their morphology, and in histochemical preparations by the presence of acid phosphatase and Oil Red 0 positive lipid droplets in the cytoplasm. The cells also contain alpha-1-antichymotrypsin and are stained by monoclonal antibody monocyte markers.

As macrophages migrate into the infarcted area, astrocytes at the periphery of the lesion undergo mitosis, hypertrophy and form large reactive astrocytes with characteristic paranuclear eosinophilic cytoplasm and multiple long thin processes. Such astrocytic proliferation begins within 1 or 2 days of the infarct and the cells are detectable by haematoxylin and eosin staining but more satisfactorily in immunocytochemical preparations for glial fibrillary acidic protein. Other techniques such as Holzer and phosphotungstic acid haematoxylin (PTAH) can also be used to identify astrocyte processes but are less satisfactory than glial fibrillary acidic protein staining.

Accompanying the astrocyte reaction, is the proliferation of endothelial cells and the extension of capillaries, together with surrounding astrocytes, into the edges of the infarct. The formation of what is, in effect, the equivalent of granulation tissue in the brain, has been quaintly referred to as a gliomesodermal reaction. A reticulin stain is particularly useful for the identification of proliferating vessels. In very small area of infarction, less than 0.5 mm in diameter, the blood vessels may grow right into the infarcted area. Frequently, it is only the demonstration of these small vessels by a reticulin technique or the associated reactive astrocytes by Holzer, PTAH or glial fibrillary acidic protein stains, that allows small infarcts to be detected in tissue sections. In larger lesions, tissue debris may remain in the centre of the infarct for weeks, months or even years and the vessels and astrocytes may only proliferate around the edge. As the debris is removed, the central area of the infarct becomes cystic and the gliosis is confined to the wall of the cyst. Macrophages may be present in such lesions for many years.

The secondary effects of an infarct may be seen in other areas of the brain. For example, if an internal capsule is damaged, the axons descending into the cerebral peduncle, pons, medulla and spinal cord degenerate. Corticospinal tract degeneration can be detected by atrophy of the tract, and loss of staining for myelin (Fig. 13).

6.3 Surgical Pathology of Tumours

Much of the immediate diagnostic service in neuropathology is concerned with the identification of brain and spinal cord tumours. Often the pieces of tissue are small and fragmented and as much assistance as possible is required from the clinical history, before attempting to diagnose the pathology of the tumour.

Prior to biopsy, the clinical history may indicate the presence of a tumour but often the exact pathology of the lesion remains uncertain. In addition to the length of history and the speed of progression of the symptoms, the age of the patient and the information derived from the CT scan offers valuable assistance to the pathologist in determining the identity of a tumour (WELLER et al. 1983). In children under 15 years of age, most tumours in the central nervous system, apart from leukaemic invasion, are primary brain tumours. Approximately 70% arise in the posterior fossa as well-differentiated astrocytomas in the pons, medulla and cerebellum or as ependymomas within the ventricles. Primitive neuroectodermal tumours (RORKE 1983) form a numerically important group in children either as medulloblastoma in the cerebellum or as cerebral neuroblastoma. From 15–30 years of age, the majority of the tumours are well-differentiated cerebral astrocytomas or oligodendrogliomas, but primitive neuroectodermal tumours in the cerebellum and cerebrum also occur in this age group. Over the age of 30 years and with a peak incidence between 60–70 years of age, anaplastic astrocytomas and glioblastoma multiforme form the majority of primary tumours, but meningiomas are not uncommon in this age group and a significant proportion of tumours biopsied are metastatic carcinomas or melanomas.

CT scans and Nuclear Magnetic Resonance (Magnetic Resonance Imaging) not only gives information regarding the position of a tumour but may also give valuable information regarding its pathology (BLEEHAN 1986). Injection of contrast medium produces enhancement in the CT scan in those areas of tumour or brain in which the blood-brain barrier is broken down (Fig. 2). As a general rule, well-differentiated astrocytomas and oligodendrogliomas do not show enhancement but they may show patchy calcification. Clinically, they usually present with a very slowly progressive history over several years often characterised by the onset of epilepsy early in the clinical course. Primary brain tumours which usually enhance include primitive neuroectodermal tumours, anaplastic astrocytomas, and glioblastoma multiforme. Characteristically, this group of tumours has a rapidly progressive clinical course of a few weeks or months. Tumours derived from tissue other than neuroectoderm also enhance; they include meningiomas and metastatic tumours. Meningiomas are usually slowly-growing tumours which present clinically once they have reached a large enough size to cause a rise in intracranial pressure or to produce localising signs from compression of adjacent brain or spinal cord. It is important to recognise these tumours as surgical removal of a meningioma is often curative. Metastatic tumours are frequently multiple and usually well-circumscribed; the prognosis, however, is usually poor and metastases are not usually biopsied unless there is difficulty in interpreting the clinical history and CT scan appearances.

Abscesses in the cerebral hemispheres often show a typical ring enhancement but may be difficult to distinguish from metastases on CT scan alone. Cerebral infarcts also enhance in the first few days and although in most cases the diagnosis can be made from the clinical history, in some instances, histology is required to establish the pathology.

Rapid diagnosis with identification of a tumour during, or shortly after, the operation is usually desirable in neurosurgery; firstly, to aprise the surgeon of the nature of the lesion upon which he is operating, and secondly to allow post operative management to be planned.

Smear or touch preparations of soft tumour tissue, or cryostat sections for firmer tissue, are ideal for the rapid diagnosis of brain tumours. One advantage of the smear technique is that smaller fragments of tumour tissue can be used than with cryostat sections. Normal brain tissue produces a smooth smear whereas tumours, and in particular poorly-differentiated tumours, result in an irregular lumpy smear due to the presence of thick blood vessels and capillary endothelial proliferation (Fig. 14) (ADAMS et al. 1981). Cytological features are well-demonstrated in smears stained with 1% Toluidine blue in 1% borax. Touch preparations can also be produced from fragments of tissue and

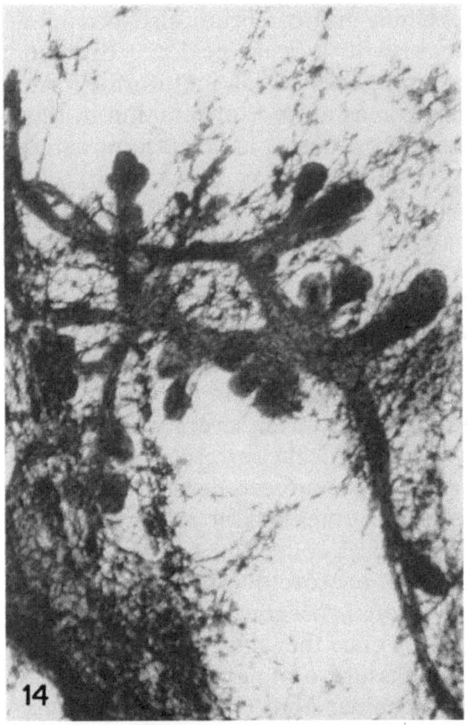

Fig. 14. Low power of smear from glioblastoma multiforme showing thick blood vessels and club-like structures formed by capillary endothelial proliferation. Toluidine blue stain, ×43

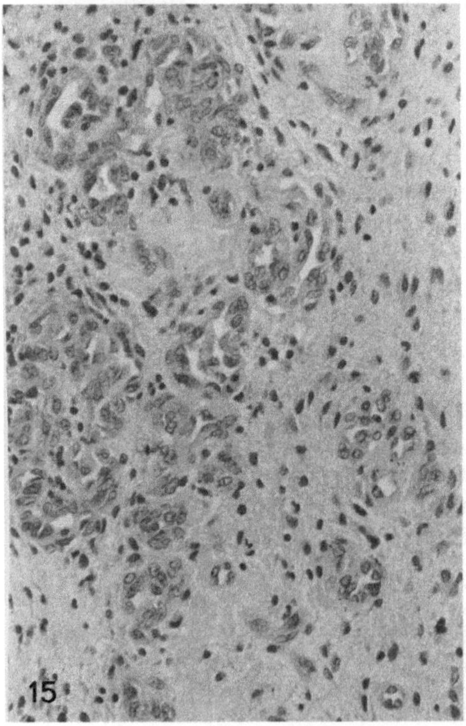

Fig. 15. Capillary endothelial proliferation in a glioblastoma multiforme. Haematoxylin and eosin, ×180

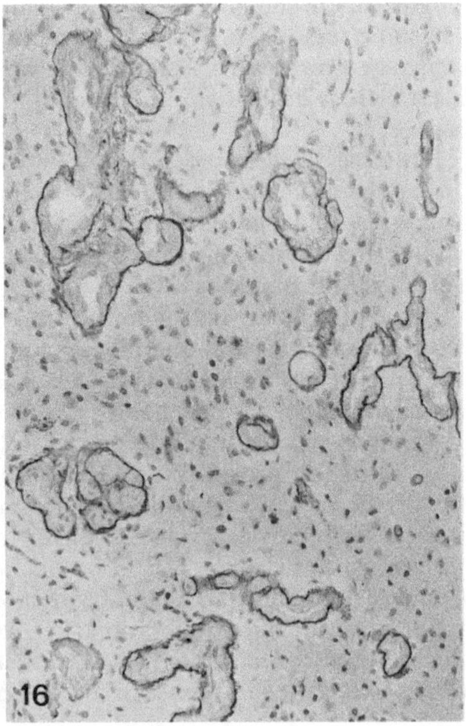

Fig. 16. Same area of tumour as Fig. 15. Vessels showing capillary endothelial proliferation are outlined by reticulin staining. Gordon and Sweet, ×150

stained with Giemsa or 'Diff-Quik'; they are particularly valuable for identifying pituitary adenomas and lymphomas.

Meningiomas, Schwannomas and some gliomas are firm and difficult to smear and cryostat sections of tissue from these tumours, quenched in liquid nitrogen, produce more satisfactory histological preparations. Cryostat sections may also be the technique of choice for the characterisation of lymphomas and diagnosis of encephalitis by immunocytochemistry (ANDERSON Chapter). In addition to Haematoxylin and eosin stains, connective tissue techniques such as Haematoxylin van Gieson and reticulin stains may be valuable for identifying tumours in cryostat sections when the cellular morphology not diagnostic. Schwannomas, for example, contain a very rich reticulin content around each cell; most meningiomas, except for the angioblastic type, however, contain little reticulin although thick bands of collagen are often present within the tumour. Most gliomas have little reticulin or collagen except around blood vessels.

It may not be possible to give a firm diagnosis of the type of tumour on a frozen section or smear but the detection of tumour and some estimate of its degree of differentiation or anaplasia may be vital information for the clinical management of the patient. Identification of most tumours is usally much more straightforward in paraffin sections. If there is a large piece of tumour available, as when a frontal lobectomy is performed or a large meningioma is extracted, care must to be taken select appropriate areas to give as much information as possible about the margins of excision of the tumour. Many of the histological features of the tumour can be evaluated in haematoxylin and eosin stained sections (Table 1) and particularly with regard to the degree of nuclear pleomorphism, nuclear cytoplasmic ratios, and mitotic indices (RUSSELL and RUBINSTEIN 1977). Glial tumours in general have a fine eosinophilic fibrillary background and the cells have indistinct borders. In addition to haematoxylin and eosin, the Haematoxylin van Gieson technique is a very useful routine stain for paraffin sections of tumours, especially for the identification of myelinated fibres in areas of infiltrated white matter. Bands of collagen are well shown in meningiomas and, in small biopsies they may be very valuable in distinguishing meningiomas from gliomas. Capillary endothelial proliferation is typically seen in glioblastoma multiforme and anaplastic astrocytomas. However, these vascular structures may not be easy to detect in haematoxylin and eosin stained sections (Fig. 15) and are much better demonstrated by reticulin techniques such as Gordon and Sweet (Fig. 16). Not all tumours form discrete localised masses in the central nervous system. Leukaemias, and some carcinomas, lymphomas and mkelanomas spread diffusely throughout the subarachnoid space and cerebrospinal fluid causing cranial nerve palsies and signs of carcinomatous meningitis which may mimic tuberculous meningitis. Such patients may also develop hydrocephalus. A diagnosis of carcinomatous meningitis often depends upon detection of tumour cells in lumbar or ventricular cerebrospinal fluid. The fluid itself must be as fresh as possible in order to preserve the cells which are most satisfactorily demonstrated in cytospin preparations stained with Giemsa or 'Diff-Quik' techniques. Immunocytochemical marker studies, particularly for the identification of lymphomas may also be performed on cytospin preparations.

6.4 Immunocytochemistry

Immunocytochemical techniques (POLAK and VAN NOORDEN 1984; STERNBERGER 1979) play two major roles in neuropathology. Firstly, they can be used as histological stains to unravel the complexities of the cytological arrangements in the central nervous system. Secondly, immunocytochemical techniques are used to specifically identify cell types within tumours particularly by the use of specific marker proteins (BONIN and RUBINSTEIN 1984; MCCOMB and BIGNER 1984) (see Table 2).

An example of the first use of immunocytochemistry is seen in Figs. 17 and 18. The pattern of invasion of the lymphoma in Fig. 17 is not clear in the Haematoxylin and eosin stained section. However, with the immunoperoxidase technique for glial fibrillary acidic protein (Fig. 18) the typical diffuse invasion of tumour cells separated into small groups by astrocytes and their processes is clearly demonstrated.

Immunocytochemistry is used in many branches of pathology for investigating the cytological origin of tumour cells. An example in neuropathology is seen in Fig. 19 in which a primitive neuroectodermal tumour contains cells which express this glial marker. Frequently, antigenic determinants, cell products or cell components are shared by several different cell types and identification of tumours in particular may require a panel of antibodies. For example, using polyclonal antibodies to S100 protein, glial fibrillary acidic protein, and Carbonic anhydrase C, it has been possible to differentiate with a reasonable degree of certainty between choroid plexus papillomas and metastatic adenocarcinomas which may otherwise have histological resemblances (WELLER et al. 1986). The choroid plexus papillomas are positive for Carbonic anhydrase C, S100 and usually focally positive for glial fibrillary acidic protein, whereas metastatic carcinomas are usually negative for all 3 markers. Other tumours such as ependymoma are positive for S100 protein and glial fibrillary acidic protein but negative for Carbonic anhydrase C. A similar system of antibody panels has been very effectively employed in the use of monoclonal antibodies in the diagnosis of cerebral tumours (COAKHAM et al. 1985). The advantages and disadvantages of some of the antibody techniques currently available for the study of the nervous system are included in Table 2. The use of monoclonal antibodies for the detection of lymphocyte subsets, macrophages and other cell elements depends upon the local availability but the range of antibodies is continually expanding and more are becoming available commercially.

6.5 Electron Microscopy

During the last 30 years, electron microscopy has been used to elucidate the fine structure of the central nervous system and the complex cellular interrelationships between neural elements (PETERS et al. 1976). In fact, no firm understanding of the morphology of the central nervous system is possible without an appreciation of the electron microscopic appearances of normal cortex, white matter and meninges. From practical diagnostic aspects, electron microscopy

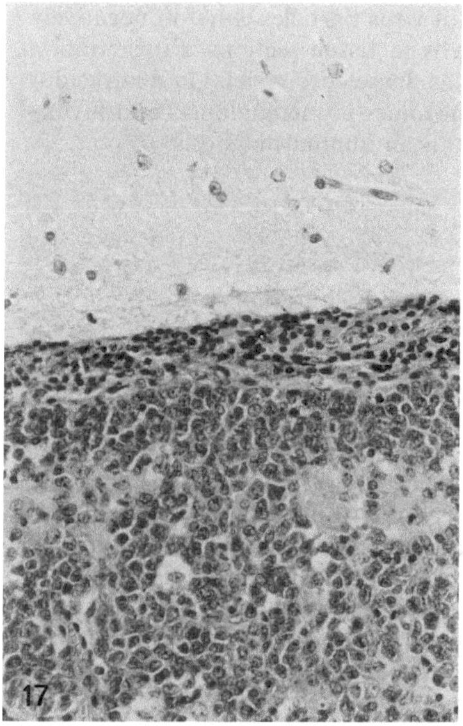

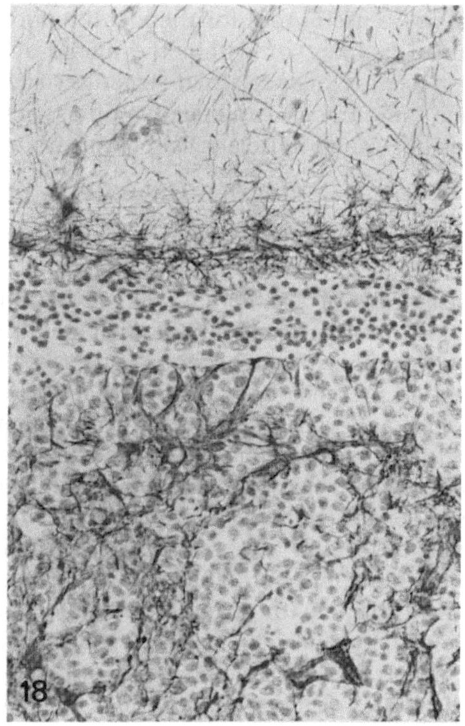

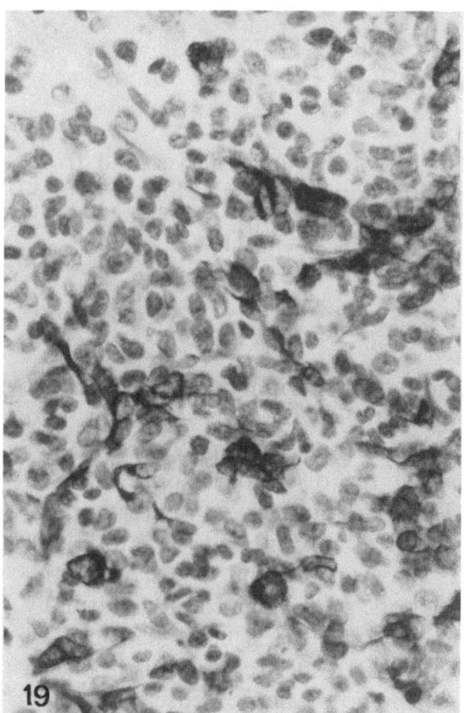

Fig. 17. Cerebral lymphoma: cerebral cortex (*top*) is separated from brain invaded by lymphoma cells (*bottom*) by a sulcus containing small reactive lymphocytes. Haematoxylin and eosin, ×195

Fig. 18. Serial section of Fig. 17. Reactive astrocytes on the surface of the cortex (*top*) are darkly stained as are the astrocytes separating groups of lymphoma cells (*bottom*). Immunoperoxidase technique for glial fibrillary acidic protein, ×195

Fig. 19. Primitive neuroectodermal tumour showing a small proportion of cells with cytoplasm stained darkly for glial fibrillary acidic protein. Immunoperoxidase stain, ×400

remains a useful tool for the identification of virus particles either in negatively stained isolates, or within the nuclei of cells in tissue sections. Furthermore, the detection of specialised structures such as dense core vesicles in neuroendocrine tumours, cilia in ependymomas, desmosomes in meningiomas and myofibrils in striated muscle cells remains important in tumour diagnosis.

6.6 Histological Artefact

The pitfalls of artefactual cyst formation and discolouration in the brain due to post-mortem artefact have been mentioned in Sect. 5.5. The histological counterpart of this artefact is poor nuclear staining, disappearance of myelin staining and the presence of basophilic bacterial colonies within the tissue. Shrinkage of cells and vacuolar artefact around neurons and oligodendrocyte nuclei are not uncommon in biopsy and autopsy material. Such changes should not be confused with hypoxic changes in neurons or with other pathological features. As a general principle, it is advisable to seek confirmatory evidence of tissue damage through the presence of inflammatory cells or astrocytic hypertrophy before ascribing pathological significance to histological changes. However, in some disorders, for example hexachlorophene and triethyltin poisoning, vacuolation of myelin may resemble the artefactual vacuolation of myelin induced by autolysis or inadequate dehydration of tissue blocks. An awareness of the possibilities of the artefact should help to avoid misinterpretation.

7 Investigation of Peripheral Nerve Pathology

Peripheral nerves arise from the brain stem to form cranial nerves and from the spinal cord to form the nerves that supply the trunk, neck and limbs. Different diseases affect peripheral nerves at different sites (Fig. 20).

7.1 Clinical Aspects

In order to adequately investigate the pathology of peripheral neuropathy, it is again important to determine the anatomical site of the disease process and its probable pathology from the clinical history and examination (WALTON 1985; WELLER et al. 1983). Most neuropathies affect the distal parts of the limbs. Some are symmetrical polyneuropathies affecting many nerves, whereas others affect just one nerve or several nerves assymetrically (mononeuritis multiplex). Dissociation of motor and sensory deficits may suggest that the neuropathy is affecting only the motor or sensory nerve roots. Motor neuropathies usually present with muscle weakness, wasting and, in some cases, fasciculation. The presence of increased reflexes and extensor plantar responses may also be seen if the upper motor neurons are involved as in motor neuron disease (amyotrophic lateral sclerosis). Loss of sensation, parasthesiae or a burning sensation

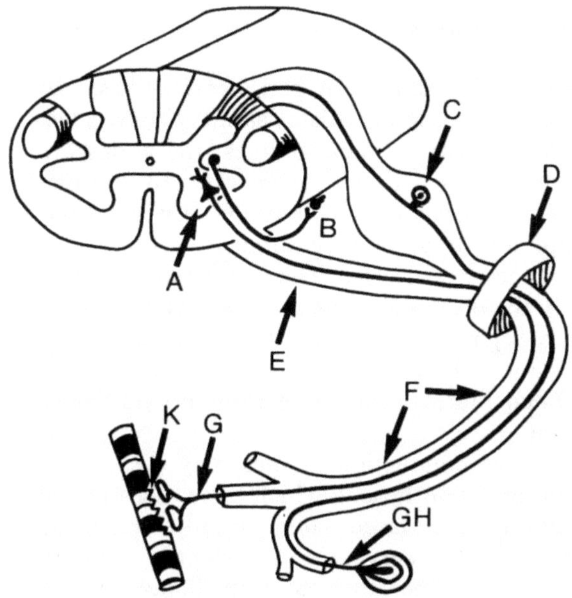

Fig. 20. Anatomical sites of damage in major disorders of peripheral nerves. (Reproduced from WELLER et al. 1983)

A, Anterior horn cells
Destructive lesions of the cord
Motor neuron disease
Spinal muscular atrophy
Poliomyelitis

B, Autonomic neuropathies
Diabetes
Amyloidosis
Familial dysautonomia
Fabry's disease

C, Sensory ganglia
Hereditary sensory neuropathies
Herpes zoster

D, Spinal canal and intervertebral foramina
Trauma
Nerve or root compression

E, Nerve roots
Post-infectious polyradiculopathy (Guillain-Barré syndrome)

F, Nerve trunks (with or without root involvement)
Trauma
Compression
Vascular disease
Demyelination in toxic neuropathies
Metabolic neuropathies
Diabetes
Amyloid
Guillain-Barré syndrome
Peroneal muscular atrophy
Hypertrophic neuropathies
Disorders of lipid metabolism
Neuropathies in malignant disease

G, Distal axonopathies (dying-back neuropathies)
Toxic neuropathies
Porphyria
Vitamin deficiencies

H, Leprosy

K, Motor end-plate disorders
Myasthenia gravis
Eaton-Lambert syndrome

in the affected areas indicate involvement of the sensory nerves. Postural hypotention, impotence, incontinence and disorders of sweating are some of the features which suggest that the autonomic nervous system is involved in the neuropathy. The time course of a neuropathy, be it acute or chronic, resolving, relapsing or progressive will all give some indication of the pathology. A significant number of rare neuropathies are hereditary so that a full family history may also be important in reaching a diagnosis.

7.2 Pathological Reactions in Peripheral Nerves

There are two main general pathological reactions in peripheral nerves, viz. axonal degeneration and segmental demyelination (THOMAS et al. 1984; WELLER and CERVOS-NAVARRO 1977). In the majority of neuropathies, the axons are damaged and the distal ends of the damaged axons degenerate. In contrast to the central nervous system, regeneration of axons is commonly seen in peripheral nerves and if distances are not too great and anatomical continuity of the nerve is maintained, regeneration may be effective. The axons remain intact in segmental demyelination and only the myelin sheaths are destroyed. Following demyelination, remyelination may begin within one or two weeks and clinical recovery may be correspondingly rapid. The commonest acute demyelinating neuropathy in Europe and the United States is post-infectious polyradiculoneuropathy (Guillain Barré Syndrome).

Clinical electromyography and nerve conduction studies may distinguish neuropathies due to axonal degeneration from those in which there is segmental demyelination. In axonal degeneration, some large axons usually remain so that maximal conduction velocities along the nerve are only slightly reduced. In segmental demyelinating neuropathies, on the other hand, the demyelination causes very significant slowing of nerve conduction velocities to fifty per cent of normal or less.

7.3 Nerve Biopsy

Nerve biopsies are effectively restricted to sensory nerves and usually to superficial nerves such as the sural nerve in the leg or the cutaneous branch of the radial nerve in the hand. Evidence of a motor neuropathy may be obtained from muscle biopsy by the detection of denervation and by the loss of axons or myelin from intramuscular nerves.

Peripheral nerves are very delicate structures and many of the problems in evaluating a peripheral nerve biopsy arise from artefact caused by rough handling at surgery and inadequate preparative techniques. The biopsy procedure (WELLER and CERVOS-NAVARRO 1977) should be discussed with a skilled surgeon and the nerve should be removed with an absolute minimum of trauma; even dabbing the nerve with a swab damages the myelin sheaths. Following careful removal, the nerve is placed on a flat piece of card and divided, if necessary, into lengths for freezing for cryostat sections, formalin fixation for paraffin embedding and glutaraldehyde fixation for resin embedding and teasing.

In the majority of biopsies, the most valuable information can be obtained from exact cross-sections of glutaraldehyde-fixed, osmium post-fixed, resin embedded nerve. In 1 µm thick Toluidine blue stained sections, the epineurial vessels can be examined for evidence of arteritis, and the degree of large and small myelinated fibre loss due to axonal degeneration can be evaluated; clusters of small regenerating nerve fibres may also be observed (Fig. 21). Segmental demyelination may be detected by the presence of large demyelinated axons

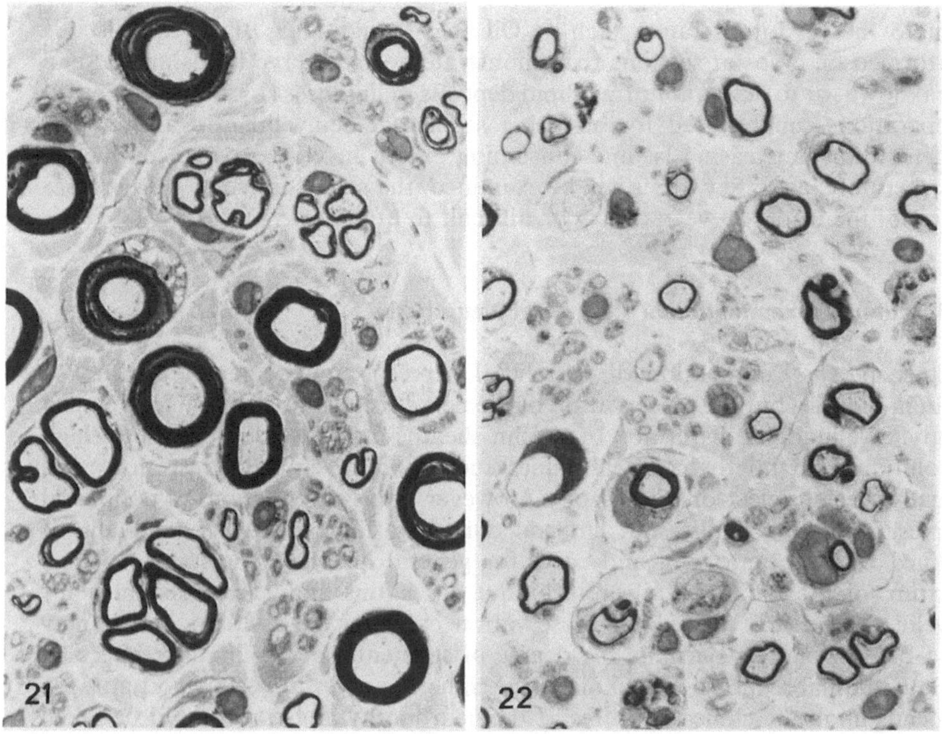

Fig. 21. Axonal degeneration and regeneration. Clusters of small, thinly-myelinated regenerating fibres are seen among the thickly myelinated large normal axons. Small non-myelinated fibres are also present. Toluidine blue stained resin section, ×1000

Fig. 22. Segmental demyelination: large demyelinated axons (e.g. *lower centre left*) and thinly myelinated remyelinating fibres are seen. Toluidine blue stained resin section, ×1000

or thinly-myelinated remyelinating axons within the nerve (Fig. 22). Onion-bulb whorls denoting hypertrophic neuropathy may be prominent in some hereditary neuropathies in which there is recurrent segmental demyelination. Such whorls have a characteristic histology which is well demonstrated in resin sections (WELLER 1984a). Non-myelinated fibres can also be assessed in resin sections by light microscopy and their numbers estimated (Figs. 21, 22).

Electron microscopy is a valuable technique for examining features within the nerve which are not resolved by light microscopy; these include the ultrastructure of myelin and the fine structure of myelinated and non-myelinated fibres.

Exact transverse and longitudinal paraffin sections of peripheral nerves stained with Haematoxylin and eosin and with Elastic van Giesen techniques are valuable for assessing damage to epineurial arteries and the presence of arteritis. Immunocytochemistry for the detection of immunoglobulins associated with myelin sheaths in the IgM kappa paraproteinaemic neuropathies (THOMAS et al. 1984) is usually most satisfactorily performed on transverse cryostat sections of nerve. Lipid stains, such as Oil Red 0 or Sudan Black can also be performed on cryostat sections to demonstrate normal nerve fibres and myelin breakdown, or to detect abnormal lipid deposits in lipidoses (LAKE 1984). Teased preparations can be used for assessing segmental demyelination and axonal degeneration but teasing is time-consuming except in very experienced hands. Furthermore, as so few fibres can be examined, the results may not be representative of the whole nerve and may be difficult to quantitate.

7.4 Autopsy Examination of Peripheral Neuropathies

The scope for examination of peripheral nerves is much greater at autopsy than in biopsies but the preservation of the nerves is less satisfactory for detailed study. Swelling and disruption of myelin sheaths may preclude the use of high resolution light microscope techniques such as resin embedded sections and usually preclude electron microscopy. However, transverse sections of peripheral nerves fixed in formalin and embedded in paraffin can be very valuable if stained with haematoxylin and eosin for general histology, or stained with a mixture of the Palmgren silver technique for axons and Luxol fast blue for myelin. The degree of axonal preservation can thus be assessed and it may also be possible to determine the degree of segmental demyelination especially in longitudinal sections. The choice of peripheral nerves to be taken at autopsy depends upon the clinical features of the neuropathy and care should be taken to label the specimens with the anatomical site of origin.

8 Muscle Diseases

8.1 Clinical Features

The general approach to diseases of muscle is very similar to that adopted for peripheral neuropathies. A clear clinical history from the patient, together with the results of the clinical examination and investigations are essential before any attempt to interpret a muscle biopsy is made. In addition, it is essential to discuss the results of the biopsy with the clinicians concerned otherwise misunderstandings may arise.

Knowledge of the age of the patient is very important. Hypotonic or "floppy" infants may have muscle disease, the commonest of which is the Werdnig-Hoffman disease with loss of anterior horn cells and extensive denervation of muscle. Alternatively, a floppy infant may have a congenital myopathy or dystrophy. Such children should be distinguished from those who are hypotonic due to perinatal brain damage. Muscle biopsy may be a decisive factor in this distinction. In older children, Duchenne dystrophy is the most devastating muscle disease and is generally manifested between 3–6 years of earlier; muscle biopsy may be required to distinguish it from childhood polymyositis. Congenital myopathies and spinal muscular atrophy also affect children. Muscle biopsy in adults may be required for the diagnosis of polymyositis, dystrophies, motor neuron disease (amyotrophic lateral sclerosis), spinal muscular atrophies and to distinguish these diseases from syndromes in which muscle weakness is a result of systemic disease.

Characteristically, primary muscle disease presents clinically with weakness and sometimes pain in the proximal muscles of the limbs, whereas neurogenic muscle disease, due to peripheral neuropathy or motor neuron disease, usually affects the distal muscles in the limbs more severely than the proximal muscles (SWASH and SCHWARTZ 1981). The time course of the disease may be an important factor in determining the diagnosis. Polymyositis, for example, may be acute, subacute or chronic in its clinical course, but most dystrophies, spinal muscular atrophies and congenital myopathies have a long slowly progressive clinical course. Dystrophies are hereditary destructive myopathies, thus the family history may be important in making the diagnosis. Once dystrophic changes have been identified in a muscle biopsy, characterisation of the dystrophy usually depends upon the clinical picture and especially the anatomical pattern of muscle involvement.

Valuable data about the pathology of muscle disease may be obtained from electromyography which can be used to distinguish between neurogenic muscle disease and myopathies. This technique is also valuable in the diagnosis of myasthenia gravis. Among the biochemical investigations, a raised Creatine kinase level in the serum indicates that muscle damage has occurred as this enzyme leaks from damaged cells into the serum. Creatine kinase levels, 10 or 20 times normal may be seen if muscle destruction is extensive as, for example, in Duchenne dystrophy.

8.2 Muscle Biopsy

The techniques applied to biopsy and autopsy specimens of muscle are very similar. Better sampling of involved muscles can be attained at autopsy, but the preservation of the muscle may not be ideal. In most cases, the diagnosis of the disease can be made by analysis of the results of a range of histochemical techniques applied to muscle (DUBOWITZ 1985; WELLER 1984a, b). Satisfactory histochemistry may be obtained on autopsy material even 24–36 hours after death. Electron microscopy is only rarely essential and then usually only for examining the structure of abnormal cellular organelles such as mitochondria.

As a with nerve biopsy, a muscle biopsy should be carefully planned and the tissue very carefully handled. The muscle chosen for biopsy should be involved clinically but not too wasted. If the disease process is too advanced, with wasting of the muscle and replacement of the muscle fibres by fibrous tissue and fat, it may not be possible to make a diagnosis from the few remaining muscle fibres. After careful exposure of the surface of the muscle, a piece for electron microscopy can be taken either by using a clamp to enclose a strip of muscle, or by stitching a 1 cm × 2 mm length of muscle to a short piece of swab stick laid on the surface of the muscle parallel to the fibres. Clamping of the muscle in this way prevents contracture when the muscle is placed in the glutaraldehyde fixative. A further piece of muscle 2 cm × 6–8 mms should be taken fresh back to the laboratory for histochemistry. Muscle samples for biochemical analyses for enzyme defects should also be taken fresh and these can be frozen directly in liquid nitrogen. The sample of tissue for histochemistry should be trimmed and orientated so that exact cross-sections can be cut in the cryostat. Care must be taken when freezing the muscle to avoid ice-crystal artefact. Some workers prefer freezing in liquid isopentane (Dubowitz 1985), others coat the muscle sample in glove powder (starch) granules, grip the muscle biopsy with a delicate pair of forceps and plunge it into liquid nitrogen (Weller 1984b). The frozen muscle can be stored in an airtight container in the cryostat for up to 24 hours, but longer periods of storage should be in a liquid nitrogen refrigerator to prevent freeze-drying.

8.3 Histological and Histochemical Techniques in the Analysis of a Muscle Biopsy

Muscle exhibits only a limited number of pathological reactions, so the diagnosis of muscle disease from a biopsy requires an analysis of the pathological changes and their correlation with the clinical features. Neurogenic muscle disease due to loss of anterior horn cells or peripheral neuropathy, is characterised by small or large groups of denervated muscle fibres and by large groups of reinnervated fibres all of the same histochemical type. In myopathies, there may be extensive fibre destruction, phagocytosis and regeneration, especially in polymyositis and dystrophies. Muscle fibre atrophy is commonly seen in myopathies but hypertrophy of muscle fibres is usually only present in dystrophies. There are a number of congenital myopathies, such as central core disease, nemaline and centronuclear myopathies in which the histological and histochemical features are so characteristic, that muscle biopsy may be the definitive diagnostic procedure. Examples of some of the common features in neurogenic and myopathic muscle biopsies are discussed in this section in order to illustrate the use of histochemical techniques in the analysis of muscle pathology.

The preservation of muscle in a good cryostat section is much better than in a paraffin section. Stained with haematoxylin and eosin, a cross-section of muscle will reveal necrosis of muscle fibres and phagocytosis together with small basophilic regenerating fibres and their large vesicular nuclei and prominent nucleoli. The presence of inflammatory cells can be confirmed by the use of an acid phosphatase technique for macrophages and by monoclonal antibody immunocytochemistry for monocyte-macrophages and lymphocyte subsets (Engel et al. 1984). If muscle fibre destruction is extensive, scarring will occur and is well demonstrated in haematoxylin van Giesen preparations.

Variation in muscle fibre size can be seen in haematoxylin and eosin stained sections (Fig. 23) but other techniques are required for characterising the pathology of this variation in fibre size (Dubowitz 1985). The adenosine triphosphatase (ATPase) technique identifies Type I and Type II fibres (Fig. 24) and is the only reliable technique for fibre typing in pathological muscle. At the standard pH 9.4, the Type II fibres stain darkly and the Type I fibres are pale.

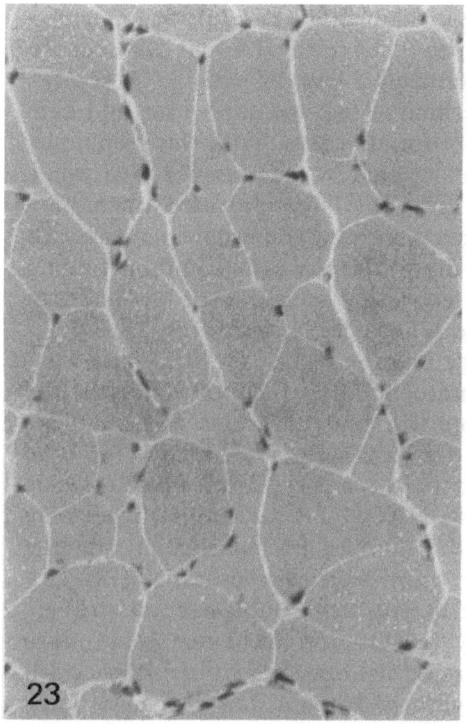

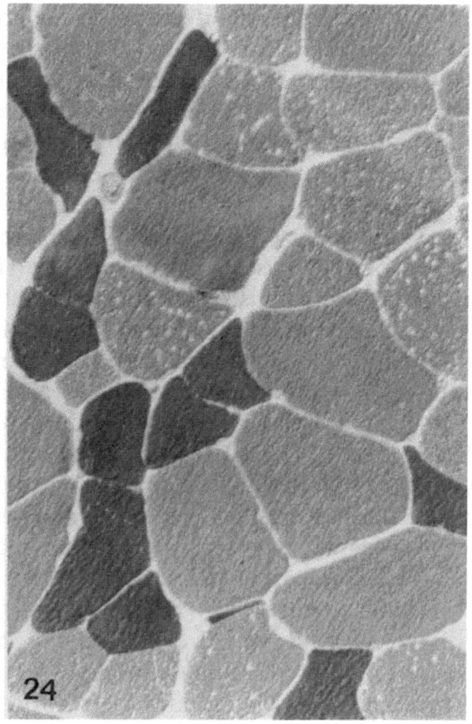

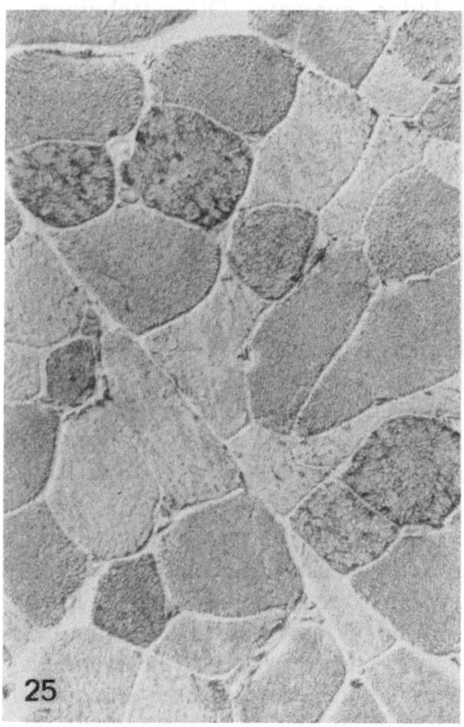

Fig. 23. Cryostat section of muscle showing variation in fibre size. Slightly darker fibres are Type 1, some of which have small holes left by extracted lipid droplets. Haematoxylin and eosin, ×300

Fig. 24. Same case as Fig. 23 showing selective atrophy of Type 2 fibres (*dark*). Holes left by lipid extraction are seen in some Type 1 fibres (*pale*). Adenosine triphosphatase pH 9.4, ×300

Fig. 25. Same case as Figs. 23 and 24. An irregular "moth eaten" appearance is seen in darkly stained Type 1 fibres. NADH-tetrazolium reductase, ×300

This pattern is reversed if the section is pre-incubated at pH 4.3 prior to incubation in the ATP mixture and 3 major types of fibre (I, IIA and IIB) can be identified if the pre-incubation is at an intermediate level of pH 4.5 (DUBOWITZ 1985; WELLER 1984a, b). The ATPase technique may be capricious and care must be taken to adjust the pH to the appropriate levels. With this technique, neurogenic atrophy within a muscle can be detected by the presence of atrophic Type I and Type II fibres; furthermore, groups of re-innervated fibres may be seen as blocks of Type I or Type II fibres disrupting the usual chequerboard pattern of the muscle. Selective Type II fibre atrophy is a common finding in patients with muscle weakness associated with systemic disorders and can only be demonstrated with certainty by the ATPase technique (Figs. 23, 24).

The nicotine adenine dinucleotide tetrazolium reductase (NADH-TR) technique stains sarcoplasmic reticulum and mitochondria. In normal muscle, the Type I (slow twitch) fibres stain dark blue with this technique and the Type II (fast twitch) fibres are lighter blue but this technique is unreliable for fibre typing in abnormal muscle. Atrophic denervated muscle fibres in neurogenic muscle disease frequently stain darkly with NADH-TR and target fibres may also be seen with this technique. In myopathic muscles, irregular patchy staining in Type I fibres produces a characteristic "moth eaten" appearance (Fig. 25). Abnormal aggregates of mitochondria or sarcoplasmic tubules stain strongly with NADH-TR and are stained red by the Gomori technique to produce "ragged red fibres". Tubular aggregates can be distinguished from mitochondrial accumulations by staining the section with the succinic dehydrogenase technique which stains mitochondria but not tubular aggregates. Other techniques may be required for the identification of specific enzyme defects in muscle (MORGAN-HUGHES 1982).

Paraffin sections of muscle may be of great value in the search for focal lesions in muscle such as sarcoid granulomata, arteritis and infarcts due to occlusive vascular disease. Electron microscopy has been valuable in establishing the ultrastructural changes in a number of muscle diseases but for diagnostic use, electron microscopy is usually limited to the identification of abnormal organelles within muscle, usually in congenital myopathies.

9 A General Approach to Neuropathological Research

Most advances in neuropathology, and indeed in medicine in general, have followed the introduction of new techniques. As the light microscope became more sophisticated in the first half of the nineteenth century, the cytology of the central and peripheral nervous systems was described by Purkinje, Remak, Schwann and many others (WELLER et al. 1983). With the introduction of osmium as a stain for myelin in the 1870's, and the use of teasing techniques, peripheral nerve fibres and their pathology were extensively investigated (RANVIER 1878). Advances in neurophysiology and the perfection of silver techniques by Golgi and Cajal in the late nineteenth century and early part of the twentieth century led to the elucidation of neural networks and to further study of axonal degeneration. During the first half of the twentieth century, developments in histological techniques using celloidin and paraffin sections allowed detailed descriptions of the pathology of many neurological diseases. The fine structure of neurons, synapses, axons and myelin was revealed by electron microscopy in the 1960's and 1970's and this lead to a greater understanding of the complexities of the central nervous system. Histochemical techniques for the study of myelin breakdown and the abnormal accumulation of lipids in lipidoses were introduced in the 1960's as were the histochemical techniques for the study of muscle pathology (DUBOWITZ 1985). During the last 2 decades, there has been an expansion of immunocytochemistry as a research tool, using both polyclonal and monoclonal antibodies (POLAK 1984; STERNBERGER 1979). Such techniques have focussed attention upon the role of intermediate filaments in cells, the distribution of neuropeptides and the heterogeneity of cell surface antigens. Work in this field continues to expand as new techniques using DNA probes appear on the horizon (EMERY 1985).

In addition to the choice of techniques, there is also another aspect to be considered in the general approach to neuropathological research. Broadly, a research project can concentrate solely upon human material, or experimental animals can be used to solve problems in human pathology. In the first category of clinical research lie the detailed descriptions of biopsy and autopsy human material using new technical developments to elucidate pathological mechanisms (ADAMS et al. 1984) or neuroanatomical structures (UPTON and WELLER 1985). In the second category, the use of experimental animals continues to be valuable in the study of basic cellular reactions and pathological mechanisms in the nervous system such as the transport of viruses (KRISTENSSON et al. 1978) and the reactions to toxic substances (SPENCER and SCHAUMBURG 1980). There is, in addition, another role for experimental neuropathology. In many diseases, it is not possible to study the basic pathological mechanisms in the human central nervous system as post-mortem material may be poorly preserved and only obtainable at an advanced stage of the disease. Experimental models have proved to be very valuable for defining the basic sequence of disease processes which can then be investigated in man. An example of this approach may be taken from the investigation of brain damage caused by hydrocephalus in neonatal and infant brains. By the use of an experimental model it was shown that in the acute high-pressure phase of hydrocephalus, the ependyma surrounding the ventricles was disrupted and the extracellular space of the periventricular white matter was expanded by cerebrospinal fluid oedema (WELLER et al. 1971). It was during this acute stage that axons in the white matter were damaged and if the hydrocephalus persisted, the periventricular white matter became gliotic. Once the sequence of changes was established in an animal model, it was possible to seek similar pathological events in isolated biopsies of human hydrocephalic brains (WELLER and SHULMAN 1972) and to use the histological appearances in such biopsies to give a prognosis for individual patients (WELLER and WILLIAMS 1975).

Finally, it should be emphasized that few concepts in biology and pathology are permanent. But, if an observation allows the formulation of a new hypothesis, this hypothesis should be tested by experimentation or further study. If the observation and the hypothesis alone are published, the work is incomplete.

Acknowledgements. I would like to thank the staff of the Neuropathology Department and Experimental Pathology for their technical assistance. I am grateful to MARGARET HARRIS for her patience in typing the manuscript.

References

Adams JH, Corsellis JAN, Duchen LW (eds) (1984) Greenfield's neuropathology, 4th edn. Edward Arnold, London

Adams JH, Graham DI, Doyle D (1981) Brain biopsy. The smear technique for neurosurgical biopsies, Chapman and Hall, London

Adams JH, Murray MF (1982) Atlas of post mortem techniques in neuropathology, Cambridge University Press, Cambridge

Bancroft D, Stevens A (eds) (1982) Theory and practice of histological techniques, 2nd edn. Churchill Livingstone, Edinburgh

Bleehan NM (ed) (1986) Tumours of the brain. Springer, Berlin Heidelberg New York

Bonin JM, Rubinstein LJ (1984) Immunocytochemistry of central nervous system tumors. J Neurosurg 60:1121–1133

Brierley JB, Graham DI (1984) Hypoxia and vascular disorders of the central nervous system. In: Adams JH, Corsellis JAN, Duchen LW (eds) Greenfield's neuropathology, 4th edn. Edward Arnold, London, pp 125–207

Bundey S (1985) Genetics and neurology, Churchill Livingstone, Edinburgh

Coakham HB, Garson JA, Allan PM, Harper EI, Brownell B, Kemshead JT, Lane EB (1985) Immunohistological diagnosis of central nervous system tumours using a monoclonal antibody panel. J Clin Pathol 38:165–173

Department of Health and Social Security (1981) Report of the advisory group on the management of patients with spongiform encephalopathy (Creutzfeldt-Jakob Disease (CJD)). HMSO, London

Department of Health and Social Security, Scottish Home and Health Department, Department of Health and Social Services, Northern Ireland and Welsh Office (1978) Code of practice for the prevention of infection in clinical laboratories and post mortem rooms. HMSO, London

Dubowitz V (1985) Muscle biopsy: A practical approach, 2nd edn. Baillière Tindall, London

Emery AEH (1985) An introduction to recombinant DNA. Wiley, Chichester

Engel AG, Arahata K (1984) Monoclonal antibody analysis of mononuclear cells in myopathies. II Phenotypes of autoinvasive cells in polymyositis and inclusion body myositis. Ann Neurol 16:209–215

Friede RL (1975) Developmental neuropathology. Springer, Berlin Heidelberg New York

Gonzalez CF, Grossman CB, Marsden JC (eds) (1985) Head and spine imaging. Wiley, New York

Kristensson K, Vahlne A, Persson LA, Lycke E (1978) Neural spread of herpes simplex virus types 1 and 2 in mice after corneal or subcutaneous (foot-pad) inoculation. J Neurol Sci 35:331–340

Lake BD (1984) Lysosomal enzyme deficiencies. In: Adams JH, Corsellis JAN, Duchen LW (eds) Greenfield's neuropathology, 4th edn. Edward Arnold, London, pp 491–572

Ludwig J (1979) Current methods of autopsy practice, 2nd edn. Saunders, Philadelphia, pp 95–129

Matthews WB, Acheson ED, Batchelor JR, Weller RO (1984) McAlpine's multiple sclerosis. Churchill Livingstone, Edinburgh

McComb RD, Bigner DD (1984) The biology of malignant gliomas – a comprehensive survey. Clin Neuropathol 3:93–106

Morgan-Hughes JA (1982) Mitochondrial myopathies. In: Mastaglia FC, Walton J (eds) Skeletal muscle pathology. Churchill Livingstone, Edinburgh, pp 309–339

Peters A, Palay S, Webster H de F (1976) The fine structure of the nervous system: the neurons and supporting cells. Saunders, Philadelphia

Polak JM, Van Noorden S (1984) An introduction to immunocytochemistry. Current techniques and problems. Royal Microscopical Society. Microscopy Handbooks II. Oxford University Press, Oxford

Ranvier L (1878) Leçons sur l'histologie du système nerveux. Savy, Paris

Rorke LB (1983) The cerebellar medulloblastoma and its relationship to primitive neuroectodermal tumors. J Neuropathol Exp Neurol 42:1–15
Russell DS, Rubinstein LJ (1977) Pathology of tumours of the nervous system, 4th edn. Arnold, London
Spencer PS, Schaumburg HH (eds) (1980) Experimental and clinical neurotoxicology. Williams and Wilkins, Baltimore
Sternberger LA (1979) Immunocytochemistry, 2nd edn. Wiley, New York
Swash M, Schwartz MS (1981) Neuromuscular diseases: a practical approach to diagnosis and management. Springer, Berlin Heidelberg New York
Thomas PK, Landon DN, King RHM (1984) Diseases of peripheral nerves. In: Adams JH, Corsellis JAN, Duchen LW (eds) Greenfield's neuropathology, 4th edn. Edward Arnold, London, pp 807–920
Upton ML, Weller RO (1985) The morphology of cerebrospinal fluid drainage pathways in human arachnoid granulations. J Neurosurg 63:867–875
Walton JN (1985) Brain's diseases of the nervous system, 9th edn. Oxford University Press, Oxford
Weller RO (1984a) Colour atlas of neuropathology. Harvey Miller and Oxford University Press, Oxford
Weller RO (1984b) Muscle biopsy in the diagnosis of muscle disease. In: Anthony PP, MacSween RNM (eds) Recent advances in histopathology 12. Churchill Livingstone, Edinburgh, pp 259–288
Weller RO, Cervós-Navarro J (1977) Pathology of peripheral nerves. Butterworth, London
Weller RO, Shulman K (1972) Infantile hydrocephalus: clinical, histological and ultrastructural study of brain damage. J Neurosurg 36:255–265
Weller RO, Steart PV, Moore IE (1986) Carbonic anhydrase C as a marker antigen in the diagnosis of choroid plexus papillomas and other tumours: An immunoperoxidase study. In: Walker MD, Thomas DGT (eds) Biology of brain tumour. Nighoff and Junk, Dordrecht, pp 115–120
Weller RO, Swash M, McLellan DL, Scholtz CL (eds) (1983) Clinical neuropathology. Springer, Berlin Heidelberg New York
Weller RO, Williams BN (1975) Cerebral biopsy in the assessment of brain damage in hydrocephalus. Arch Dis Child 10:763–768
Weller RO, Wisniewski H, Shulman K, Terry RD (1971) Experimental hydrocephalus in young dogs: histological and ultrastructural study of the brain tissue damage. J Neuropathol Exp Neurol 30:613–627

Dementia in Middle and Late Life

C.L. SCHOLTZ

1 The Nature of Dementia: Definition

A major problem in the study of dementia has been the production of a definition. That most commonly used is that of LISHMAN (1978): – an acquired global impairment of *intellect*, *memory* and *personality*, but without impairment of consciousness. This definition is important because it identifies the clinical nature of the problem and directs study to those areas of the brain which have a major part to play in these functions. It also excludes patients with clouding of consciousness (delerium) as a major symptom, or when the confused state is secondary to toxic and infective intracranial or extracranial disease.

Confusion may be associated with dementia. One quarter of the elderly in hospital show some confusion and one third of demented patients have superadded confusion. This change may be associated with treatable conditions (such as heart failure or infection) and thus full clinico-pathological correlation is essential in the study of these patients. Involvement of the sensory receptive areas in pathological processes may also contribute to confusion.

Dementia is a spectrum of disease ranging in severity from those mildly demented, clinically little different from benign senescent forgetfulness, to those with severe dementia. A dementia score is necessary to assess patients fully and to make comparisons. Scores are also useful in assessing progression which is not always the rule, only one third of patients with mild dementia progress to the severe form of disease. In any patient mental function can vary considerably and an individual may be able to function to an apparently high level because of social support. The disease processes involved may result in a short or long course (acute or chronic brain failure). All of this variability emphasises the value of a scoring system.

Some patients with a significant neurological defect may be identified earlier in the course of the disease than those in which dementia is the main presenting problem, thus an individual who is apparently intellectually normal may be examined pathologically after death from unrelated causes and be found to have the changes associated with a severe dementing illness. It is only on close questioning of the family that the limited capacity to perform daily functions, the lack of correct social skills and reduced control of emotional reactions are determined retrospectively.

The pathological processes in dementia may extend beyond the areas related to intellect, memory and personality as in Huntington's chorea (basal ganglia) and Parkinsonism (nigro-striatal tract).

The pathological substrate of dementia is destruction of one or more areas of the brain involved with intellect, memory (particularly short term) and personality. The disease processes which effect these usually destroy the cortex, especially in the frontal lobe, the temporal lobes and several associated subcortical areas. It is becoming increasingly recognised that several areas are commonly involved in any one disease. In general terms the disease processes which produce dementia produce comparable clinical syndromes. Alzheimer's disease affects all areas that are associated with dementia; other diseases affect these areas to a greater or lesser extent and may, in addition, involve other areas producing specific neurological defects.

Despite extensive study there is still a not inconsiderable group of demented individuals for whom no pathological explanation of the dementia can be given (KIM et al. 1981).

2 The Problem

Dementia in middle (41–65) and late adult life (over 65 years) presents a major social problem. The elderly population is increasing and so are the numbers of those with dementia. The prevalence of dementia has been estimated at 2% in the 65–70 year old group and 20% in those over 80 (Report of Royal College of Physicians 1981). Of these about half have a severe dementing illness. However because of the limitations of psychological assessment and diagnostic procedures this may be an under estimation. There appears to be no sex preference (Tomlinson et al. 1970). Individuals over 90 appear to be resistant to the development of Alzheimer's disease and cerebral atherosclerosis (Larsson et al. 1963).

The cause of dementia in middle and late life varies as many of the dementing diseases of middle life cause death before 65. The incidence of atherosclerosis and its CNS manifestations also increase progressively with advancing years. In late adult life Alzheimer's disease and multi-infarct dementia may coexist and in a small group of patients the entity producing the major component of the dementia may be difficult to assess.

The dementing illnesses in middle life are a combination of those of early life, in which the individuals have survived into middle life, and those seen only in middle life. In the former group Downs syndrome is important because of dementia associated with Alzheimer type changes in the brain. In the latter group Creutzfeldt Jacob disease has a short course so patients die in this age range. Similarly alcoholic dementia, which shortens the life span of the individual, and trauma are mainly problems of middle life. However, further confusion may be produced by the diseases of late life presenting prematurely. The systemic diseases which produce dementia span the two groups and reflect the disease patterns of the community.

The epidemiology of dementing illnesses is poorly understood. Many studies do not have adequate pathological back-up (Marsden and Harrison 1972) and when they do there is a 25% discordance between clinical diagnosis and pathological findings (Corsellis 1962).

In toxic and metabolic disorders dementia is important and the neuropathology is not well described. The major changes are generally non-specific in type and secondary to complications such as anoxia, epilepsy or hypotension. Most are uncommon and seen in specific situations, for example, where there is gross vitamin deficiency (pellagra) or occur uncommonly in association with other disease (pernicious anaemia). The importance of this group is that many patients present with dementia, particularly myxoedema and respond well to treatment.

3 The Study, History, Autopsy and Histological Technique

The clinical study of demented patients must be accompanied by a full history with particular note of drugs taken or recent drug withdrawl, mental examination using a dementia score and laboratory studies including a brain scan. The history may suggest the areas of the central nervous system to which attention should be directed. A full autopsy is necessary to separate patients with dementia due to generalised disease from those with generalised neurological diseases and intracranial pathology. The autopsy should include the dissection of the neck vessels and removal of the spinal cord including some posterior root ganglia along with representative samples of nerve and muscle. The dissection of the cerebral vessels should be undertaken in the fresh specimen to identify berry aneurysms around the circle of Willis as these are difficult to identify in the fixed brain. Where viral disease is suspected samples of the cortex, particularly the temporal lobe, should be taken immediately after removal both for electron microscopic examination and culture. Biochemical study of neurotransmitters is possible when the brain is available for study shortly after death.

The brain weight is one of the most important observations to record. External examination of the cortex should follow, preferably with the leptomeninges removed, to reveal possible tissue loss. The cranial nerves should also be identified and the cerebellum weighed separately after fixation. Fixation of the intact brain in buffered formal saline is important as slicing of the unfixed brain with subsequent fixation results in distortion of tissue and difficulty in the selection of blocks. The method used to section the brain varies with local practice. The traditional method of coronal slices starting at the mamillary bodies and proceeding anteriorly and posteriorly has been partially superseded by cutting slices to match those of the brain scan. The features to note are the ventricular size, size of deep grey matter and abnormalities in the white matter and cortex.

In Picks disease focal atrophy of the frontal and temporal lobes is highly suggestive on external examination. Atrophy of the corpus striatum reflected by enlargement of the lateral ventricles in relation to the head of the Caudate nucleus in the sliced brain is diagnostic of Huntingtons Chorea. In several disease processes full assessment can only be made with extensive study of all areas particularly in late life where there is a combination of age and disease.

The selection of blocks depends to a great extent on the disease thought to be most likely to have produced the dementia. In most instances all areas of the cortex should be sampled. These include the deep white matter. In addition the hippocampus, blocks of the deep grey matter, brain stem to include the substantia nigra and cerebellum are necessary.

The special stains which are the most useful are a stain for central myelin – luxol fast blue with a counter stain to identify nissl substance such as cresyl fast violet; a stain for glia – the immunoperoxidase technique for Glial Fibrillary Acid Protein (GFAP) has become widely available and has some advantage for a routine laboratory over the more traditional Holzer method. There are many silver techniques available for axons and the choice depends on local expertise. The Palmgren technique, particularly the Cross modification (Cross 1982) or the modified Bielschowsky (Yamamoto and Hirano 1986), have been recognised as the best for identifying neurofibrillary tangles. Histochemical and immunohistological methods which include methods for the study of neurotransmitters are becoming commercially available at reasonable cost. Electron microscopy studies of subcellular structures and possible aetiological factors in infective dementing disease are limited because fresh material is limited to the occasional biopsy and autopsy samples. Newer computer systems have allowed counting of large numbers of neurons in an accurate and reproducible manner to characterise the changes which occur with age and identify pathological processes.

4 Classification of Dementia

The pathological classification of dementia has been unrewarding. The earliest method separating those where the pathology was mainly in the cortex from those where it is in the subcortex has not stood the test of time. Alzheimer's Disease, for example, involves both cortical and subcortical areas. The two commonest entities, Alzheimer's Disease and multi-infarct dementia, which have separate aetiologies and were once thought not to occur together, have now been described in association. The pathological process seen in Alzheimer's disease are common to those seen in old age and are thus commonly seen in elderly individuals with a dementia of vascular origin. Thus a broadly based classification which includes both macroscopic and microscopic findings is the easiest to present but it must be remembered that CNS diseases can occur in association and be complicated by extracranial pathology.

Patients with temporary confusion and patients with primary memory disorders, particularly those with Herpes simplex encephalitis where the brunt of the disease falls on the temporal lobes, should not be regarded as demented and so are excluded from this group. As with all clinico-pathological studies the separation of individuals with benign senescent forgetfulness from the group of demented patients in a study can be difficult. It is only with full antemortem assessment using a precise definition of dementia that a valid decision can be made.

5 Alzheimer's Disease

Alzheimer's disease is the commonest cause of severe dementia in middle and late life. In the "presenile group" (under 65) dementia typically presents between 50–60 with death occuring two to three years after diagnosis (Wolstenholme and O'Connor 1970). Clinically the presenile form is similar to the senile form except that focal and parietal symptoms may be evident and the pathology may be more severe. A family history of Alzheimer's disease has been reported in up to one third of cases (Tomlinson and Corsellis 1984) and many have a family history of psychiatric illness. The pathology in these cases is identical to that seen in the sporadic cases.

5.1 Macroscopic Findings

The major macroscopic observation made in this disease is reduction in brain weight resulting from loss of cortex and white matter. However, brain weight may rarely be normal (Terry 1980). In intellectually normal individuals there is a progressive reduction in brain weight with age starting at 20 and accelerating

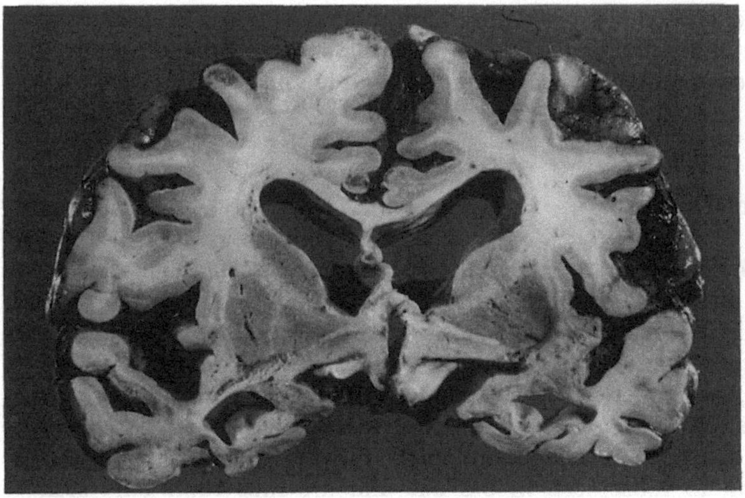

Fig. 1. Coronal slices from the brain of a 60-year-old severely demented patient with Alzheimer's disease

after 50 in an exponential manner (DEKABAN and SADOWSKY 1978; DAVIS and WRIGHT 1977). In Alzheimer's disease cortical loss occurs before white matter loss (CORSELLIS 1962, 1976) affecting both frontal and temporal lobes (Fig. 1) rather than the occipital. The apparent loss of cortical tissue is probably secondary to shortening (DUYCKAERTS et al. 1985), rather than thinning (TERRY 1980). There is secondary loss of white matter with enlargement of ventricles reflecting loss of large neurons in patients less than 80 years old (HUBBARD and ANDERSON 1985) and in normal aged individuals (BRODY 1955; COLON 1972; TERRY 1980). When the brain is severely shrunken minimal trauma can cause subdural haematoma, and intracerebral haemorrhage can occur from vessels severely weakened by congophilic angiopathy (MANDYBUR 1986). The complications of atherosclerosis described in the section on vascular disease can also occur in aged individuals with Alzheimer's disease although the younger cases tend to have less severe disease than intellectually normal individuals because of their poor nutrition.

5.2 Neurofibrillary Tangles

Neurofibrillary tangles (Fig. 3) were first described by Alzheimer nearly 80 years ago (WILKINS and BRODY 1969) and can be identified in both intellectually normal and Alzheimer's disease patients. They are intraneuronal dense bundles of filaments which may fill the neuronal body, extending around the nucleus displacing the organells without disrupting them. They extend into the axon and are best seen using the Cross modification of the Palmgren silver impregnation method (Fig. 2) (CROSS 1982) or by the production of a green birefringence in polarised light when stained with congo red due to the β pleated structure. Electron microscopy shows them to the bundles of paired helical filaments

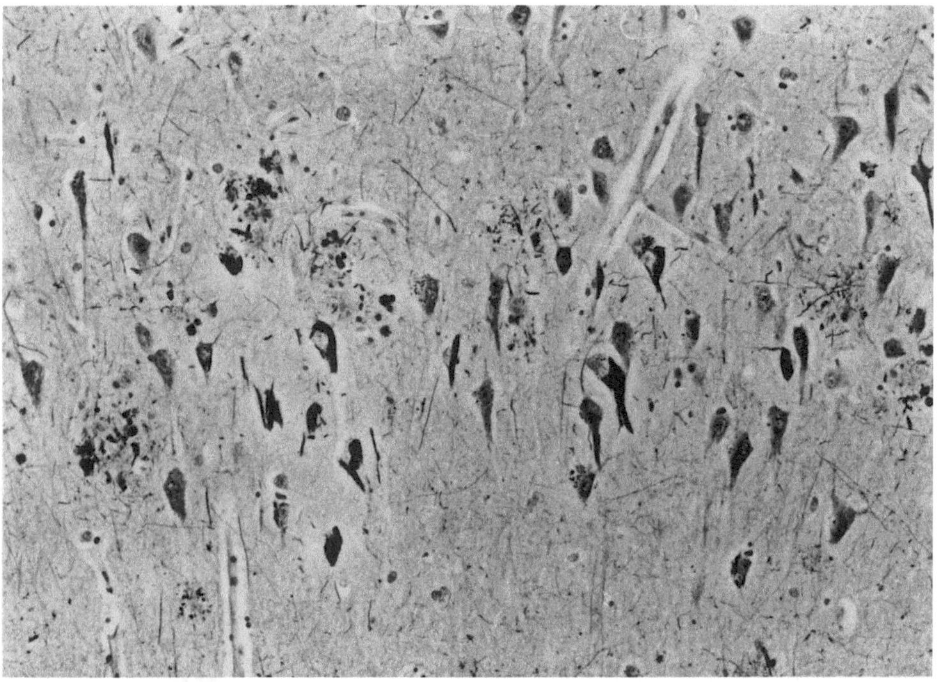

Fig. 2. Photomicrograph of section from the temporal lobe showing large numbers of neuritic plaques and neurofibrillary tangles. Modified Palmgren, ×150

(KIDD 1964) (Fig. 3b) each about 10 nm in diameter with a helical twist every 80 nm (WISNIEWSKI et al. 1972). When the neuron disintegrates they remain as dense bundles. Neurofibrillary tangles are derived from neurofilaments and belong to the group of intermediate filaments. Neurofilaments are formed in the cell body and move distally within the axon existing as dispersed subunits or aggregates dispersed within the cytoplasm. There physical state is affected by ionic conditions in the cytoplasm and drugs, e.g. Colchicine and Vinblastine (WEISENBERG and TIMASHEFF 1970).

The neurofibrillary tangle may result from an abnormality of synthesis of the underlying subunits or proliferation of the filaments secondary to injury to the terminal synapses and axons. Microtubules can be disrupted experimentally by elevation of serum calcium causing filament aggregates to form (WISNIEWSKI et al. 1964) or may accumulate following axon section in association with vincristine neuropathy (SHELANSKI and WISNIEWSKI 1969). Thus the primary event may be the development of the neuritic plaques with secondary development of neurofibrillary tangles, a possibility supported by the earlier appearance of plaques. Aluminium can also cause the accumulation of neurofilaments (KLATZO et al. 1965), however these are unpaired 10 nm filaments. Increased amounts of aluminium have been found in neurons of Alzheimer's disease bound to DNA (CRAPPER et al. 1976) and may have a role in their development in Alzheimer's disease. Full characterisation of neurofibrillary tangles has not been achieved. The tangles are highly insoluble which has made

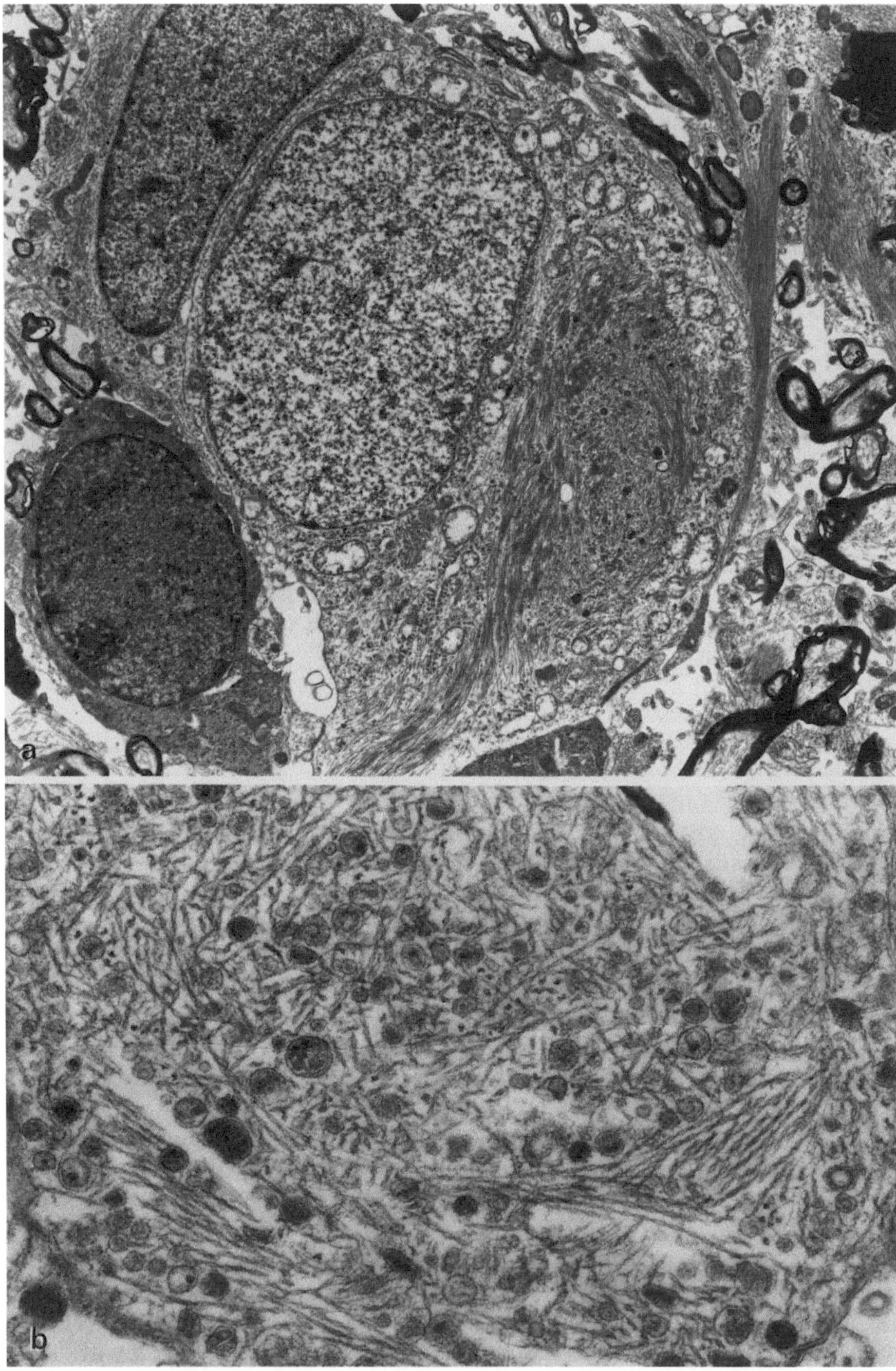

Fig. 3. a Cortical neuron in Alzheimer's disease to show intracytoplasmic neurofibrillary tangles. Note displacement of the nucleus and subcellular organelles in the cytoplasm. EM Mag ×5738. **b** Detail of bipaired twisted tubules in longitudinal and in cross section. Some tendency to group in parallel bundles can be observed in lower half of the picture. EM Mag ×26400. (Courtesy of Dr. Luis H. Carrasco)

biochemical analysis and the development of antibodies particularly difficult. Biochemical study using electrophoresis has shown that in addition to neurofilaments there is a polypeptide molecule of about 50000 daltons (IQBAL et al. 1974, 1982a, b) possibly related to tubulin. Preliminary immunocytochemical study suggests that the Alzheimer neurofibrillary tangles are different from normal neurofilaments and tubules (IQBAL et al. 1982b) particularly in relation to the distribution of the phosphorylated neurofilament protein (CORK et al. 1986). It is not possible to say if the tangles in the demented and the non demented are the same and the contribution of other parts of the cytoskeleton is uncertain.

The number of tangle bearing neurons in the neocortex correlates well with the degree of dementia regardless of the patients age (WILCOCK and ESIRI 1982). The tangles are most evident in large neurons and can be correlated with the reduction in their number (TERRY et al. 1981; HUBBARD and ANDERSON 1985). Tangle bearing neurons are found in the upper and lower layers of the cortex in the presenile form and predominantly in layer three in the senile form (DUYCKAERTS et al. 1985) with relative sparing of layer four. There seems to be a columnar arrangement of the neurons most affected (PEARSON et al. 1985) an observation which correlated well with the apparent shortening of the cortex (DUYCKAERTS et al. 1985) suggesting that there is selective damage to individual columns.

5.3 Neuritic Plaques

Neuritic plaques (Figs. 2, 4, 5) are composed of abnormal neuronal processes many of which are presynaptic terminals containing mitochondria, glial cells and amyloid. They are evident in the third decade before neurofibrillary tangles in intellectually normal individuals (TOMLINSON et al. 1970; TOMLINSON and HENDERSON 1976) and increase rapidly in number in the fifth and sixth decade.

They were first described by FULLER (1911) and their importance both in ageing and Alzheimer's disease was recognised by BLESSED et al. (1968) and TOMLINSON et al. 1970). The thickening of neurites seen in the early stage of plaque formation may result from interference with axonal flow. The plaque matures with the release of amyloid into the centre (TERRY and WISNIEWSKI 1970), the amyloid having probably formed locally from catabolised antibody – antigen complex released from blood vessels (WISNIEWSKI et al. 1975). Macrophages are present in the centre of plaques and there is an increase in Cathepsin A, used as a marker for brain macrophages in the cortex of patients with Alzheimer's disease (BOWEN et al. 1977). The plaque also contains reactive astrocytes. Plaques can develop experimentally in the absence of amyloid in animals treated with aluminium (WISNIEWSKY and TERRY 1968) and so amyloid may not necessarily be the initiating factor.

Since a correlation of dementia score and the number of neuritic plaques by BLESSED et al. (1968) a critical evaluation of this work has lead to doubt being expressed on the relevance of neuritic plaques to the dementia. This is because of a lack of homogeneity in the demented patients in the later study

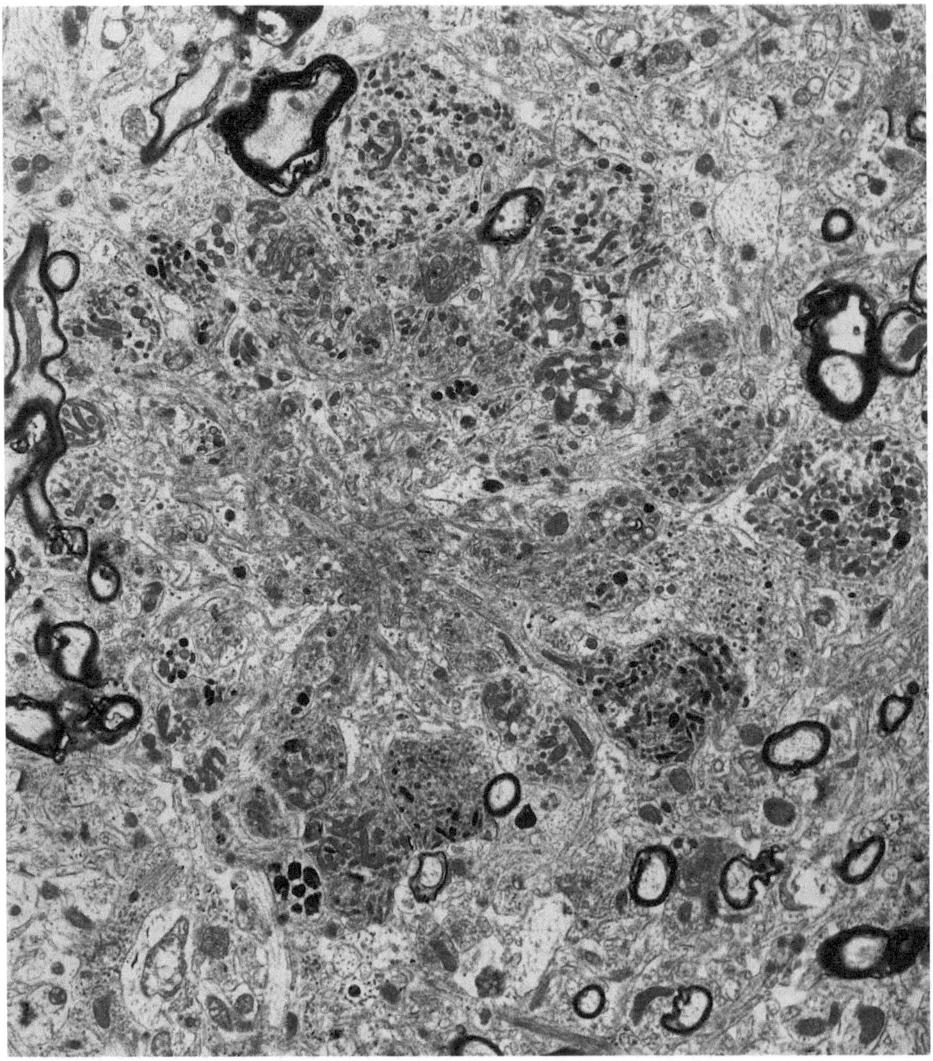

Fig. 4. Neuritic plaque in cerebral cortex in Alzheimer's disease. Swollen cell processes can be seen in the margins of the plaque containing dense, membranous bodies and numerous mitochondria. Note displacement of myelinated axons. EM Mag ×7545. (Courtesy of Dr. LUIS H. CARRASCO)

(BALL 1978). Additional work has indicated that there is a better correlation of dementia with neurofibrillary tangles (WILCOCK and ESIRI 1982). The number of neuritic plaques and neurofibrillary tangles does not seem to be interdependent as some cases have numerous plaques and few tangles and vice versa.

The cognitive impairment in Alzheimer's disease can also be correlated with the cholinergic deficit in the cortex and changes in the ascending cholinergic system (PERRY et al. 1978).

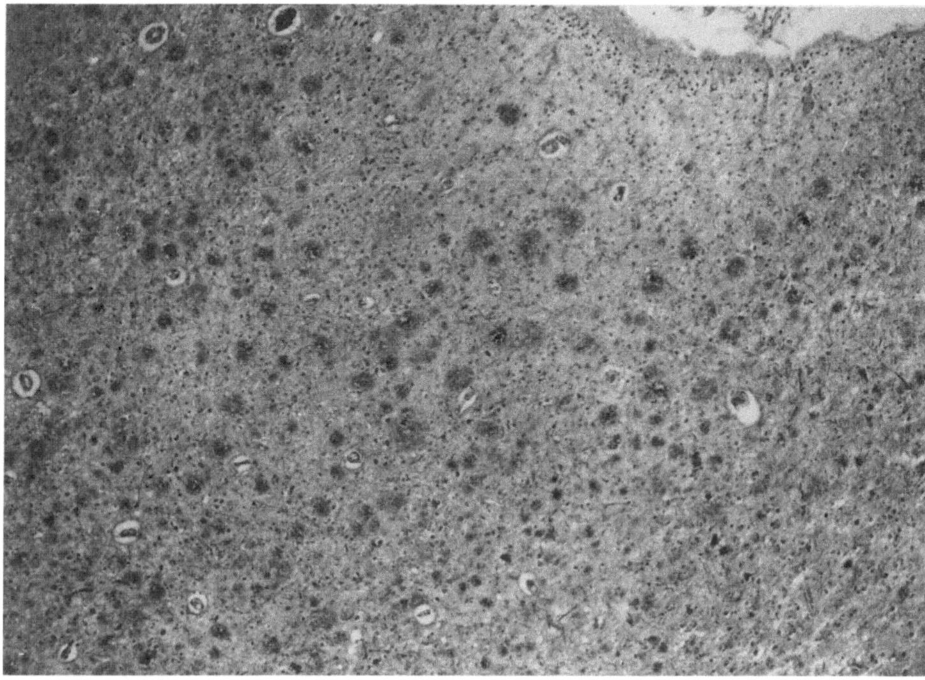

Fig. 5. Low power photomicrograph of section of the cortex from an intellectually normal individual showing large numbers of neuritic plaques. Palmgren, ×44

5.4 Hirano Bodies and Granulovacuolar Degeneration

The two other important correlates of normal ageing and Alzheimer's disease are Hirano bodies (Fig. 6) and granulovacuolar degeneration of neurons (Fig. 7). Hirano bodies were first described in the Parkinsonism dementia complex of Guam (HIRANO 1965) and are present in the temporal lobes both adjacent to and away from the hippocampal pyramidal neurons. They appear as paracrystalline ovoid or spheroid eosinophilic bodies on light microscopy. Electron microscopic study shows that they are sheets of membrane bound ribosomal particles derived from partially degraded rough endoplasmic reticulum (O'BRIEN et al. 1980). Hippocampal neurons also contain numerous vacuoles with membrane bound inclusions from 3–5 mm in diameter described as granulovacuolar degeneration. The granules have an irregular electron dense core surrounded by electron lucent material.

5.5 Amyloid

The incidence of amyloid deposition in vessels increases progressively with age (BEHAN and BEHAN 1979). Amyloid may be found (predominantly in meningeal vessels) in normal aged individuals. The term congophilic angiopathy is applied

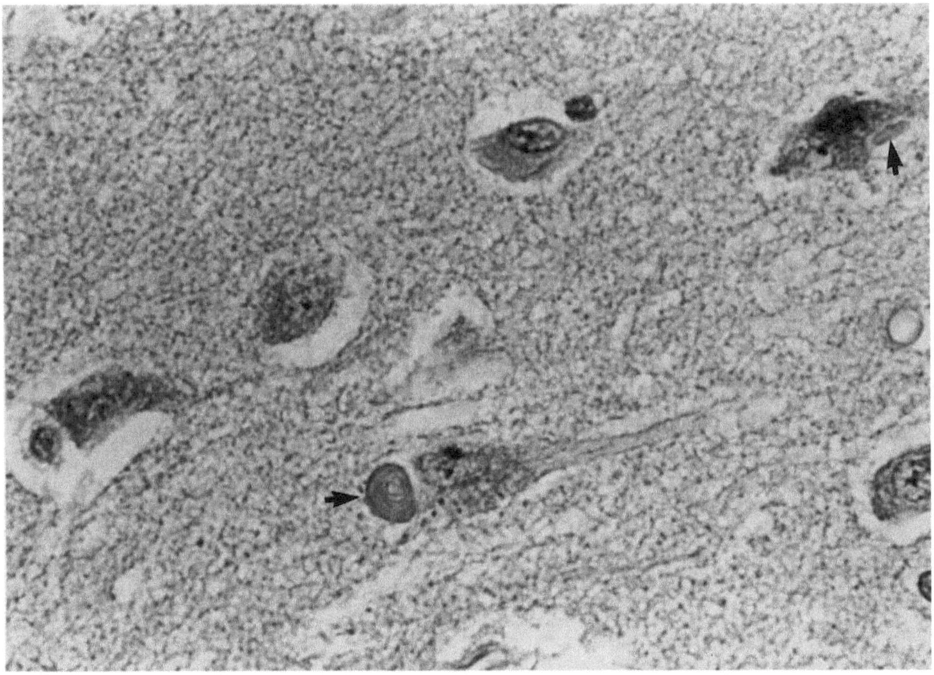

Fig. 6. Photomicrograph of a section from the temporal lobe of a patient with Alzheimer's disease showing severe extracellular paracrystalline Hirano bodies (*arrows*). H&E, ×87

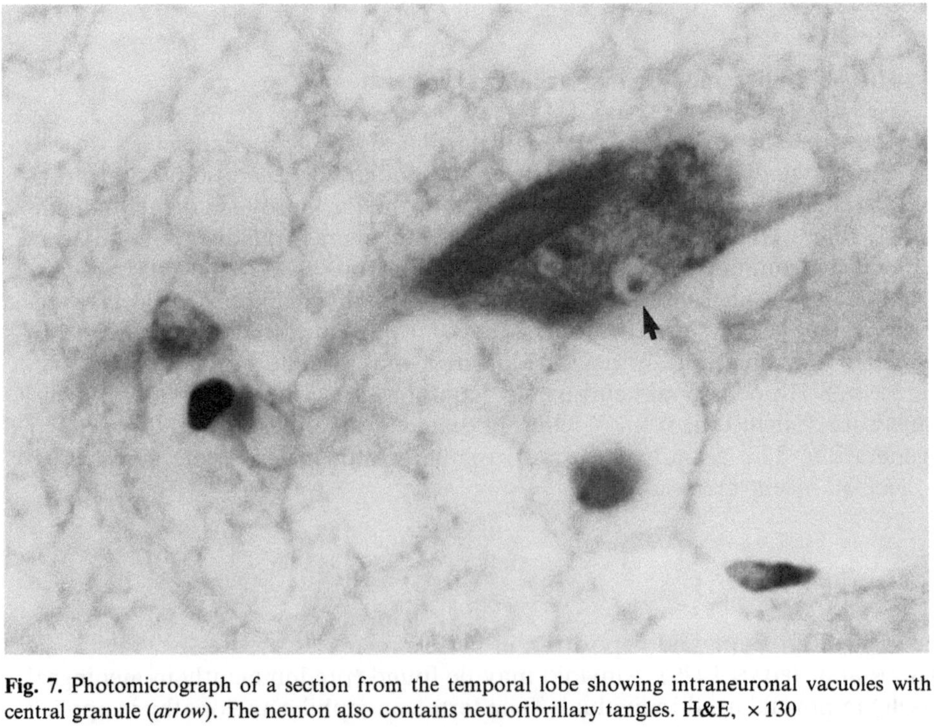

Fig. 7. Photomicrograph of a section from the temporal lobe showing intraneuronal vacuoles with central granule (*arrow*). The neuron also contains neurofibrillary tangles. H&E, ×130

when amyloid is deposited in cerebral arterioles, capillaries and vessels (SCHOLZ 1938). Amyloid may extend into the surrounding brain which, if associated with intellectual impairment, constitutes a unique form of dementia (TORACK 1978). Cerebrovascular amyloid is one of the commonest findings in Alzheimer's disease (MANDYBUR 1975, 1986). Amyloid is initially observed in the muscularis and adventitia. The amyloid may be formed in the vessel wall (GLENNER et al. 1981), and the resulting damage may allow other amyloidogenic protein to leak into the brain to be converted to amyloid in the plaque. Plaque amyloid contains less tryptophan and tyrosine than perivascular amyloid which suggests that local factors may be important in its production (POWERS and SPICER 1977).

Amyloid in vessels and plaques shares several physiochemical properties with neurofibrillary tangles. Both have a fibrillar structure and a β pleated configuration which is birefringent on congo red staining. These properties are shared by the prion (PRUSINER et al. 1983), a protein particle which does not contain DNA or RNA, and is the infectious agent in scrapie (MCKINLEY et al. 1983). PRUSINER et al. (1983) suggested that amyloid may be the infectious agent in Alzheimer's disease. Scrapie prions may be the infectious agent in Creutzfeldt-Jacob disease, a "spongyform encephalopathy" which is histologically dissimilar to Alzheimer's disease. Alzheimer's disease appears to lack transmissibility to animals which may mean that the animals used are not permissive hosts.

5.6 Neuronal Pathology

Considerable technical difficulty has been experienced in counting neurons in the brain in both intellectually normal and demented individuals. Most recent studies have shown a small but significant loss of neurons with increasing age (BRODY 1955; TERRY 1980) and in Alzheimer's disease (HUBBARD and ANDERSON 1985) particularly in the younger patients. The numbers of neurons in the brain stem are stable in both groups. The most significant cellular abnormality is loss of dendritic spines in the pyramidal nerve cells throughout life most marked in layer three pyramidal cells of the prefrontal and superior temporal cortex (SCHEIBEL and SCHEIBEL 1975). The basilar dendrites are lost first followed by the branches until there is a knobbly apical dendrite with few branches. These authors suggested that these changes are secondary to neurofibrillary tangles in Alzheimer's disease.

The accumulation of lipofusion is easily observed in many neuronal groups with age. It represents hydropic indigestable material which probably represents the accumulation of cellular debris and is representative of the level of cellular activity (SCHOLTZ and BROWN 1978). There is no evidence that it is deleterious to the cell or that it is present in excessive amounts in Alzheimer's disease (MANN and SINCLAIR 1978).

Studies of the cortical neurotransmitter particularly the intrinsic cortical neuropeptide somatostatin show a significant reduction (DAVIES et al. 1980; BEAL et al. 1985) which appears to be selective to the frontal and temporal cortex. A subclass of somatostatin positive neurons may be affected selectively in Alzheimer's Disease (ROBERTS et al. 1985). Somatostatin immunoreactive processes are present in neuritic plaques which are most commonly found in layer three. This suggests that plaques form by degeneration of presynaptic and postsynaptic neurites of large layer three or five projection neurons which form the cortical association areas (MORRISON et al. 1985). In addition there are reduced amounts of the biosynthetic enzyme for acetylcholine – choline acetyltransferase (CAT) (ROSSOR et al. 1982) in the cortex. The origin of the ascending CAT neurons projecting to the cortex is in the basal forebrain. The subcortical nuclei which comprise this complex are the medial septal nucleus, the nucleus of the diagonal band of Broca and the basal nucleus of Meynert. All three are cholinergic and project to the cortex, including the hippocampus. The loss of this cholinergic activity in the cortex has been attributed to the loss of neurons in the nucleus of Meynert. This observation led to the development of the cholinergic hypothesis whereby the loss of the cholinergic input to the cortex is seen as the major determinant of the pathogenic process (HARRISON 1986). Similarly the reduced activity of the noradrenergic marker enzyme dopamine-beta-hydroxylase activity in the frontal and temporal cortices and hippocampus is associated with loss of cells from the locus ceruleus (CROSS et al. 1981) and the dorsal raphe nucleus projects serotonin to the neocortex and reduced amount of serotonin has been reported in the cortex (BENTON et al. 1982). However physical damage to the cortex results in shrinkage of neurons in the nucleus basalis of Meynert (SOFRENIEW et al. 1983). This suggest that the subcortical pathology is secondary to the cortical degeneration. The somatostatin changes could be explained by the pathology in the cortex.

5.7 The Amygdala

The amygdala, a subcortical structure, may also play an important role in the pathogenesis of Alzheimer's disease. Many of its functions – emotion, motivation and memory – are disturbed in Alzheimer's disease. Bilateral lesions of the amygdala in animals produce striking abnormalities of bahaviour and affect (DICKS et al. 1969). The anygdala has extensive, direct and reciprocal connections with the frontal cortex (NAUTA 1961). In Alzheimer's disease there is a 70% decrease in the absolute number of neurons in the amygdala (HERZOG and KEMPER 1980) and it is a selective loss affecting the medial nucleus and the medial portion of the central nucleus.

5.8 Isodendritic Core

The subcortical nuclei all contain a large number of neurons with a non-specialised isodendritic pattern of extensive intermingling with other neurons. These form a continuous core extending from the spinal cord to the basal forebrain. These neurons have been grouped under the heading "isodendritic core" and are similar to what was previously called the reticular formation. The pathology seen in these nuclei is firstly neuronal loss which is uniform throughout the nuclei and which is correlated with reduction in nucleolar volume. This suggests that there is a progressive reduction in protein synthesis leading to cell death (MANN et al. 1984). In the locus ceruleus the neuronal loss is not closely correlated with the clinical and pathological features of dementia (TOMLINSON et al. 1981), and the neuronal loss in the nucleus of Meynert of elderly patients is greater than that of demented cases of Alzheimer's disease (PÉRRY et al. 1982).

Patients with parkinsonism can present with a cognitive defect of varying severity. Where the dementia is severe neuropathological study shows all the features of Alzheimer's disease and the two diseases appear to coexist. There is also a group of patients with parkinsonism who have a mild dementia and do not show the pathology of Alzheimer's disease in the cortex. In this group there is a markedly reduced cortical CAT level (PERRY et al. 1983). The dementia in both Parkinson's disease (WHITEHOUSE et al. 1983) and Alzheimer's disease is related to cortical cholinergic defects and to loss of neurons in the nucleus of Meynert. Neuronal loss is no greater in Alzheimer's disease despite the more severe dementia. This suggests that the cognitive impairment in Alzheimers disease originates from a primary cortical abnormality with secondary changes in the ascending cholinergic system, this supports the concept that the cognitive impairment is not necessarily a function of neuronal loss in this disease. It does not exclude the possibility that other functional defects are also present, as the intellectual impairment correlates more closely with cortical neurochemistry than changes in the subcortical nuclei. The common pathology of the neurons in the isodendritic core suggests that they are selectively vulnerable in this disease process.

5.9 Conclusion

Whether the primary pathology in Alzheimer's disease is the cortical degeneration or the changes in the subcortical nuclei has yet to be decided. The readily apparent cortical pathology suggests that this is the initial process and the common anatomy of the subcortical nuclei suggests that they may be destroyed because they are selectively vulnerable. These observations do not exclude the possibility that other neuronal groups are involved.

At present there are no absolute pathological criteria for the diagnosis of Alzheimer's disease. Alzheimer patients may have a brain of normal weight with minimal cortical pathology and intellectually normal individuals may have large numbers of neuritic plaques with significant numbers of neurofibrillary tangles. Future study must separate these by quantitating the pathological features they share and by studying the pathological process early in the disease.

5.10 Aetiology

The study of the aetiology of Alzheimer's disease is primarily the search for an environmental agent which after interacting with the genetic background of the individual produces the disease. The identification of the agent has been slowed by the lack of an experimental model. The neurofibrillary degeneration produced by aluminium toxicity is not the same as the neurofibrillary tangles seen in Alzheimer's disease the distribution of plaques around blood vessels suggest that a toxic agent is important. Small amounts of silicon have also been identified in plaques which may also be significant in the formation of plaques because of the formation of aluminosilicates (Candy et al. 1986).

The immune system and the role of the T-cell in the disease process has been investigated with particular regard to its role in the development of amyloid. The neuropathological changes of Alzheimers disease are found in most patients with Down's syndrome dying over the age of 35 years. Abnormalities in the nucleus basalis of Meynert (Casanova et al. 1985) and thus a possible genetic factor may be important in Alzheimer's disease.

6 Dementia and Vascular Disease

The arterial supply of the head and neck can be separated into the large muscular arteries where atherosclerosis has a major impact and the small intracerebral vessels with a diameter of less than 2 mm where several pathological processes can produce secondary effects on the brain. Atherosclerosis of the extra- and intracranial muscular arteries results in narrowing of the lumen of these vessels but it is the thromboembolic complications (Blackwood et al. 1969) which lead to large cerebral infarcts or multiple infarcts of varying size with or without dementia. Only rarely does cerebral infarction result from in situ thrombosis (Lhermitte et al. 1970) and is then usually secondary to a vasculitis or compression of an artery. The term atherosclerotic dementia has been criticised because the dementia results from loss of brain tissue and not from atherosclerosis; demented and non-demented individuals may have atherosclerosis to a similar extent (Corsellis and Evans 1965). A major association of atherosclerosis is hypertension and clinical association with dementia is common (Landurner et al. 1982a). Other factors include diabetes, cardiac disease and hyperviscosity of the blood which may contribute to the development of vascular lesions. However these risk factors are not seen more commonly in demented individuals than in those who have similar vascular disease and are non-demented (Landurner et al. 1982a).

6.1 Multi-infarct Dementia

Occlusion of large vessel results in infarction of a selected area dictated by the field of supply of the vessel. The lesion usually includes both grey and white matter and the neurological defect produced is related to the area destroyed. Mental impairment occurs when sufficient tissue has been lost from those cortical areas related to intellect, memory and personality, particularly the frontal and temporal lobes, however bilateral posterior cerebral infarction

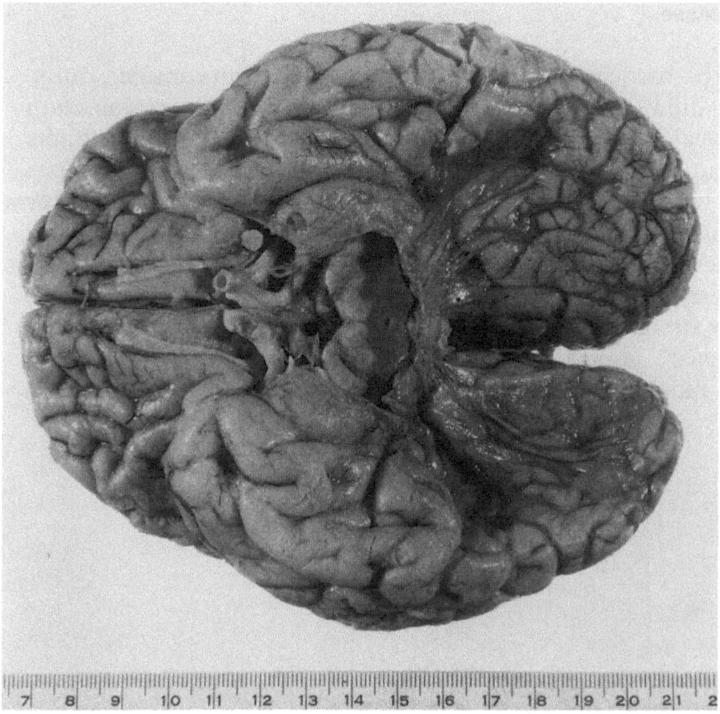

Fig. 8. Photograph of the base of the brain from an 82-year-old demented woman showing bilateral infarction of the temporal and parietal lobes

can also result in severe intellectual impairment (Fig. 8). In addition the damage to the white matter also produces effective decortication, which has a similar effect. Destruction of those areas specifically involved with sensory perception may add confusion to the intellectual impairment. This is particularly true when the individual is cortically blind.

The total volume of tissue lost in demented individuals is generally in excess of 100 ml and intellectual loss can be directly related to volume (BLESSED et al. 1968). Dementia can rarely result from a single infarct when the dominant hemisphere is involved but typically the infarcts are bilateral; the basal ganglia, thalamus, (LANDURNER et al. 1982b) hippocampus and corpus callosum are more commonly the site of an additional infarcts in demented patients than in non-demented individuals. The term "multi-infarct" dementia has been applied to this group of patients (HACHINSKI et al. 1974) and the typical patient with multi-infarct dementia has had several episodes of cerebral infarction with improvement between episodes, with some retention of insight until sufficient tissue is lost and dementia ensures. The true incidence of multi-infarct dementia is not known. The low incidences of 17% (TOMLINSON et al. 1970) was regarded as more accurate because neuro-psychological disfunction was excluded from the definition of dementia.

6.2 Binswanger's Disease

Dementia may also be associated with loss of the deep white matter. BINSWANGER (1894) first drew attention to this entity. Macroscopic examination of the brain may reveal little except dilated ventricles (Fig. 9). Small infarcts may be evident, "lacunes", secondary to occlusion of long perforating arteries (FISHER 1965, 1969) with considerable loss of white matter (Fig. 9). However microscopic examination of the white matter shows thickening of the small blood vessels with widening of the spaces around them. These small intracranial vessels with a diameter of less than 2 mm can be affected by hypertension, microaneurysm formation, amyloid deposition and microembolism. These pathological processes are associated with infarct like changes in the deep white matter and may, in a manner that is not fully understood, produce a dementing

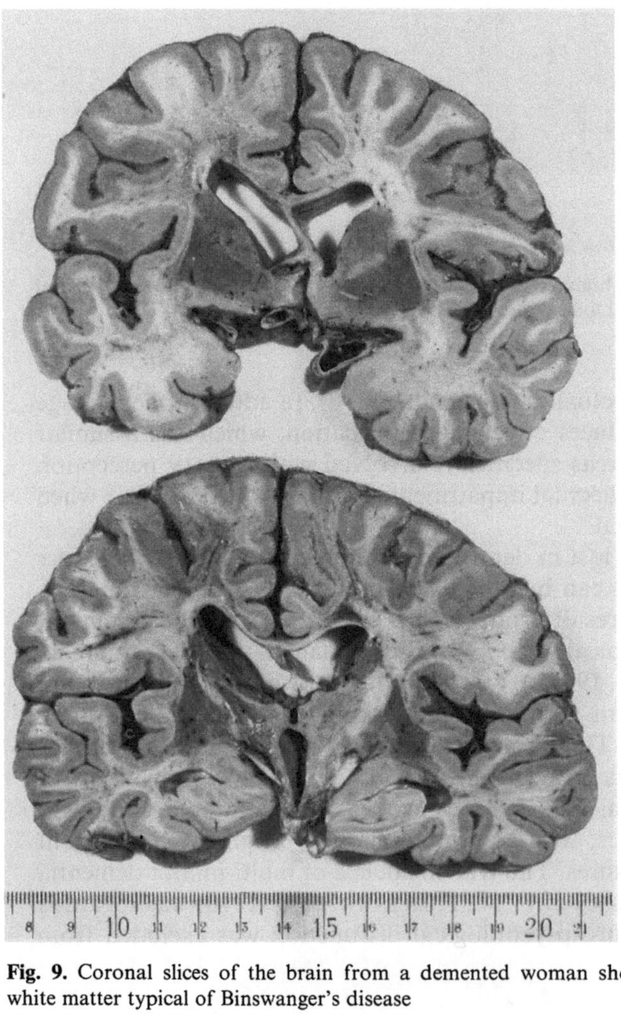

Fig. 9. Coronal slices of the brain from a demented woman showing extensive infarction of the white matter typical of Binswanger's disease

illness. The process is typically associated with hypertension and the resultant pathological appearances are similar to those seen in infarction of the white matter with gliosis, cavitation, and loss of white matter with iron filled macrophages around blood vessels (JANOTA 1981). The cortex is generally spared as are the subcortical U fibres (DE REUCK et al. 1980) and there are usually no large infarcts (JANOTA 1981). These patients typically are younger than the patients with Alzheimer's disease and the dementia is not as profound. It may be complicated by a complex neuropsychiatric state.

6.3 Boundary Zone Infarction

Ischaemic brain damage along cerebral arterial boundary zones may complicate profound hypotension particularly in association with myocardial infarction. The period of hypotension necessary to cause infarction appears to be in excess of seven minutes (MALONE et al. 1981). Hypoxia causes infarction in the fields between the anterior and middle and middle and posterior cerebral arteries. Minimal lesions are usually confined to the parieto-occipital lobes which involve the visual areas. This sequence of events is most common in elderly individuals and so may be seen in individuals with pre-existing dementia and may further complicate the clinical features.

Dementia may be associated with berry aneurysms. These are rarely large enough to produce a space occupying effect, with sufficient tissue loss to cause dementia, but a large anterior communicating aneurysm may destroy sufficient tissue in the frontal lobes for this to occur. The more likely sequence is rupture with spasm of the anterior cerebral arteries with or without obliteration of the subarachnoid space by blood. The sequence of events following this are tissue loss due to infarction and hydrocephalus secondary to failure to absorb CSF.

7 Dementia and Infection

Dementing illness following infection is usually due to virus infection. They may be grouped into those that produce dementia *acutely* following infection with disorders of intellect, memory and personality immediately apparent, and those which have a *subacute or chronic* course with symptoms developing over months or years. In the former group the CNS pathology may be secondary to *direct invasion* of the brain by the virus or to a *post-infective* process where the primary infection occurs outside the CNS. This group includes measles, mumps, varicella, rubella, smallpox and vaccinia. This group of infection and sequelae most commonly affects individuals in childhood and early adult life.

7.1 Acute Infection

The commonest cause of sporadic acute encephalitis in western Europe is Herpes simplex encephalitis. Many patients die in the acute episode of the illness survivors include patients who have a profound memory defect with only minor intellectual and personality loss. One third have focal neurological lesions. Patients who die years after the onset of the disease show the long term effects of necrosis of one or both temporal lobes, particularly the inferior and middle temporal gyri and the anterior parts of parahippocampal gyrus. The hippocampus, amygdaloid and putamen with related white matter are commonly involved and show marked cystic degeneration.

The group of viruses where extra CNS infection produces perivenous demyelination includes a large group of which measles has been the best studied. The neurological sequelae develop 5 to 25 days after the primary infection and mortality can be up to 10% with 1% of survivors having persistent neurological defects. These can include abnormalities of cognition. The common events in all these infections appears to be an attack on myelin by sensitised lymphocytes.

7.2 Subacute and Chronic Infection

Measles can also be associated with a subacute illness described by Greenfield (1950) as *subacute sclerosing panencephalitis*. The age range of presentation of this disease is from 2–21 years it and runs a course of several years of behavioural change and mental deterioration progressing to coma and death. The large number of intranuclear inclusion initially described by Dawson (1933) suggested

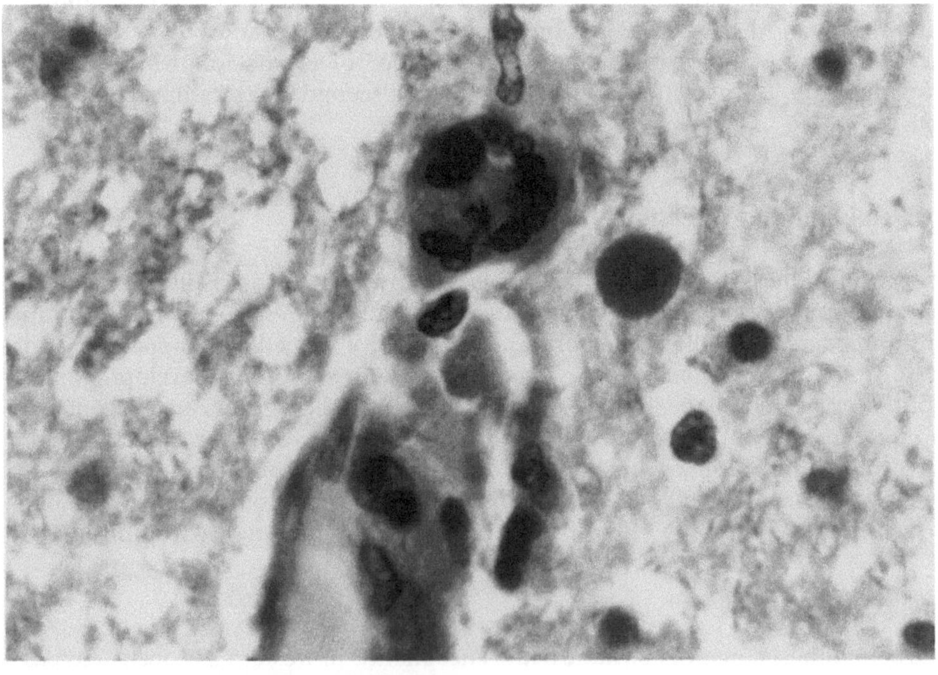

Fig. 10. Multinucleate giant cells adjacent to a blood vessel from a patient who died of AIDS. H&E, ×210

a viral aetiology. They were identified as virus by electronmicroscopy by Bouteille et al. (1965).

The human immunodeficiency virus (HIV) persistently infects a small minority of T-helper lymphocytes in the peripheral blood, lymph nodes and spleen; it also persistently infects cells throughout the brain (Shaw et al. 1985). In the brain it slowly duplicates and can be detected as integrated and unintegrated viral DNA and persistently expresses the infectious RNA virion. The cytopathic effect on the brain is cumulative and becomes symptomatic around five years after infection. It can occur without severe immune deficiency (Siegal 1985) or opportunistic infection. It is characterised by the presence of multinucleated giant cells around blood vessels in the white matter (Fig. 10) and are similar to those seen in lymph nodes (Sharer et al. 1985).

Individuals suffering from a disease which affects their immune competence (lymphomas, leukaemia and carcinomas) or are on immunosuppressive therapy can develop a dementing illness with weakness, visual loss and ataxia. This was first described by Astrosom et al. (1958) as *progressive multifocal lucoencephalopathy* and is characterised by focal areas of necrosis and demyelination (Fig. 11) with large bizarre oligodendroglial cells in the inflammatory reaction

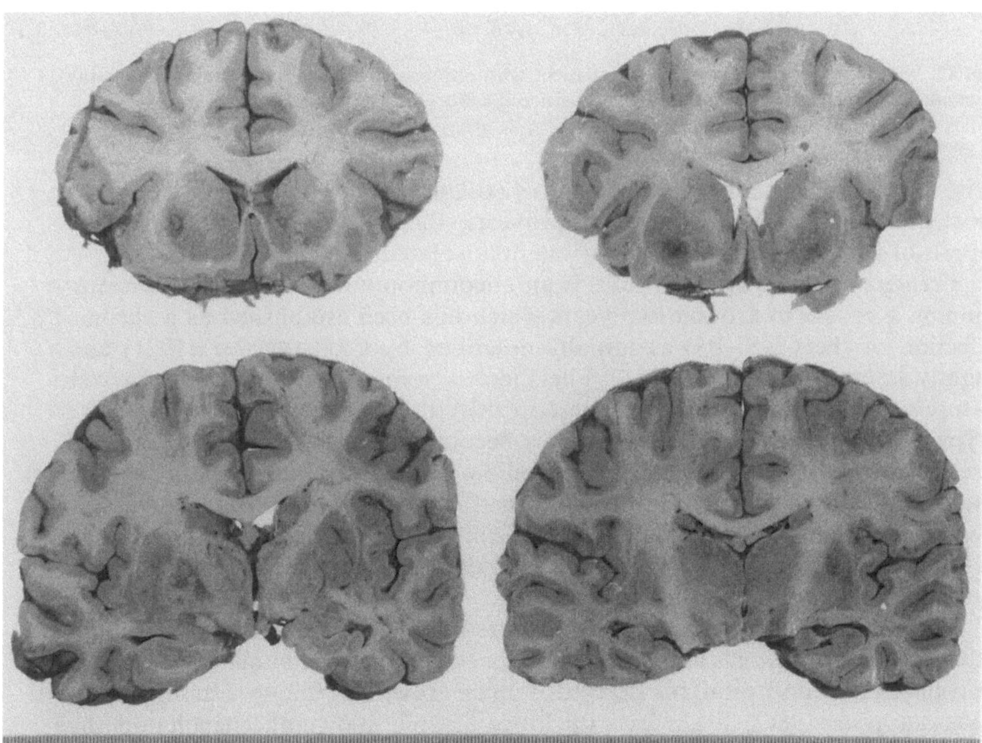

Fig. 11. Coronal slices of the brain showing focal areas of necrosis in a patient with progressive multifocal leucoencephalopathy

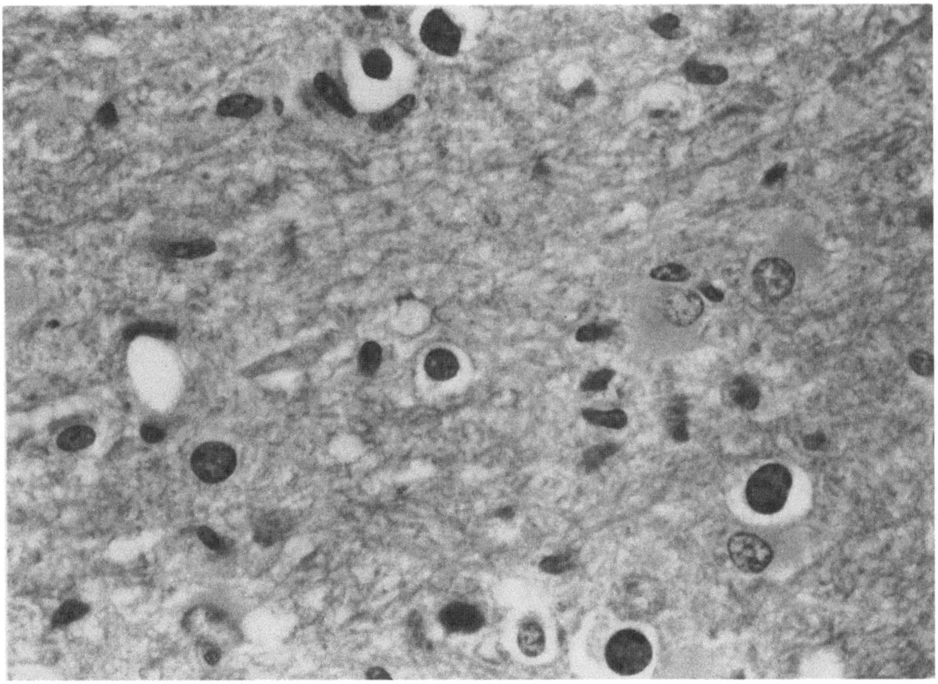

Fig. 12. Photomicrograph of section from patient with progressive multifocal leucoencephalopathy showing numerous bizarre oligodendroglial nuclei and astrocytes. H&E, ×172

(Fig. 12). These contain viral inclusions (ZURHEIN and CHOU 1965). The virus most commonly identified is a papovavirus, the J C virus named after the initials of the patient from whom it was first isolated (PADGETT et al. 1971).

Creutzfeldt-Jacob disease (CJD) is an uncommon human presenile dementia running a course of around two years which has been established as a chronic infection of the CNS. It was initially described by CREUTZFELDT (1921) and slightly later by JAKOB (1921). While there is some debate as to the precise nature of the disease in the cases described by these authors, which were very atypical, the existence of this entity has become established. The incidence of the disease is about one case per two million. It is disease of middle life which usually present with a prodrome of forgetfulness and confusion progressing over several weeks to a profound dementia. In addition there may be pyramidal, extrapyramidal and cerebellar signs. A typical EEG pattern of periodic sharp wave complexes supports the diagnosis.

The disease has been readily transmitted experimentally to primates and other laboratory animals and accidentally passaged between humans via corneal transplantation and pituitary derived human growth hormone administration (WELLER et al. 1986). It shares with kuru, scrapie and mink encephalopathy the production of a status spongiosis of the cortex and deep grey matter, a spongy appearance in histological sections being secondary to neuronal loss and astrocytic proliferation (Fig. 13).

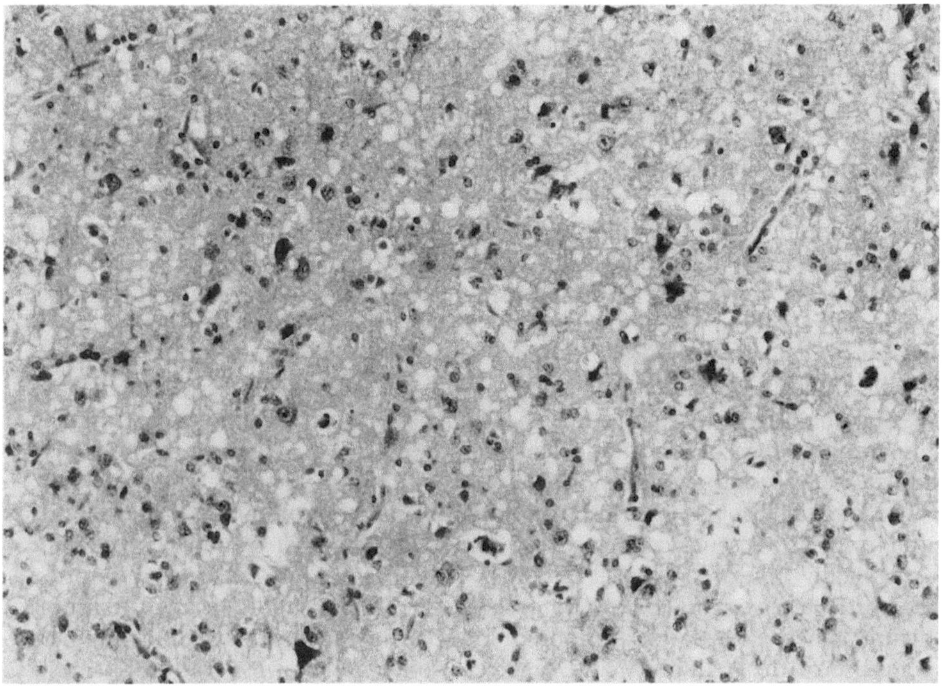

Fig. 13. Photomicrograph of section of a cortical biopsy from a patient with Creutzfeldt-Jacob disease showing "spongyform change", shrunken neurons and astrocytes. H&E, ×43

The mean age of death of sporadic CJD is 57.8±9.7 years and in the familial form 51.1±9.6 years while in the Gerstmann-Straussler syndrome, which may include variants of the Creutzfeldt-Jacob disease the average age at death is 48.0±10 years, (MASTERS et al. 1981a, b). Sporadic and familial cases of CJD have a similar mean age at death. In contrast cases secondary to iatrogenic transmission, usually present well outside these age ranges. This suggests that contact with the transmissible infective agent in the environment is not required for the onset of either the familial or sporadic cases. The term "virogene" was suggested by LWOFF (1960) to described a gene sequence, present in the host, which may under certain circumstances when the gene sequence is not suppressed code for an agent which may replicate or induce changes in the cell surface (RIDLEY et al. 1985). RIDLEY et al. postulated that when the endogenous gene sequence is not suppressed there may be expression of age dependent mechanisms which control the onset of deleterious familial neurodegenerative disorders. This may include some cases of CJD and Gerstmann-Straussler syndrome.

Bacterial and Protozoal infection of the CNS may also produce a dementing illness due to destruction of the brain or obliteration of the subarachnoid space with secondary hydrocephalus. Parenchymal destruction of the brain by syphilis – general paresis of the insane – is rare. Dementia presents late and insideously and sometimes may be accompanied by features of tabes dorsalis. Pathologically the brunt of the disease falls on the frontal lobes where there is neuronal loss and astrocytic proliferation. The pathology is secondary to tissue necrosis due to invasion by the treponema and infarction secondary to endarteritis obliterans.

8 Trauma and Dementia

In England and Wales 210 totally disabled individuals survive head injury and 1500 are left severely disabled each year (LEWIN et al. 1979). The outcome of head injury can be reliably predicted using the Glasgow coma scale, identifying the age of the patient, the worst state of neurological responsiveness and duration of post traumatic amnesia. Patients who survive with decerebrate dementia die within one year. Those who survive with severe neurological sequeli have a near normal life expectancy while those with a brain stem, cerebellar or minor hemiparetic pattern of lesion and leave hospital have a reduction of life expectancy of five years (LEWIN et al. 1979).

The mental disability varies from a mild post traumatic syndrome mainly characterised by resentment, to severe dementia. The post traumatic syndrome is organically determined (KELLY 1981) and is a major cause of disruption of domestic and working life. The more severe mental disability is a frontolimbic dementia where severe impairment of day to day memory occurs and there is a release of the irritability and rage reaction together with frontal dysmesia and frontal personality change in some instances. The intellectual function tests are more impaired when the trauma is right sided (SMITH 1974).

The tissue damage in a closed head injury is thus secondary to the energy transmitted into the skull and the intracranial structures. Skull fractures also reflect the amount of energy transmitted to the skull but are commonly absent where there is severe disruption of the brain with coma. The sequeli of a closed head injury are focal or diffuse. Focal lesions are cortical contusions, intracranial haematoma and damage secondary to raised intracranial pressure, while diffuse damage is generally evident in the white matter (diffuse axonal injury), blood vessels and also apparent as diffuse hypoxia and swelling.

The damage to the cortex consists of haemorrhagic necrosis and oedema on the tips of the gyri. In time these become brown and depressed. The lesions are typically distributed over the tip and orbital surface of the frontal and temporal lobes and around the sylvian fissure. The contusion index (ADAMS et al. 1980) derived from the degree and extent of the contusions, shows that contusions are more severe in patients who have a fracture of the skull and are less severe in patients who have white matter lesions (diffuse axonal injury). This is reflected in the observation that there is no difference in the severity of contusions in patients who do or do not loose consciousness at the time of head injury. Boxers, who experience recurrent trauma to the head commonly show small pigmented areas on the frontal and temporal lobes. Histologically the neurons in the cortex contain neurofibrillary tangles in these individuals (CORSELLIS 1973) in the absence of neuritic plaques.

The damage to the white matter is the most important part of the immediate effect of head injury, producing loss of consciousness and subsequent dementia.

It is seen when the applied acceleration is prolonged (GENARELLI et al. 1982) and results from tearing of axons. It can be seen diffusely through the hemispheres and focally in the corpus callosum (Fig. 14a, b), and where the impact is sufficient, in the dorsolateral quadrant of the rostral brain stem (MITCHELL and ADAMS 1973). The tearing results from the shearing forces in a tissue with a low capacity to resist shearing deformation but a high resistance to compression (HOLBOURN 1943, 1945). The appearances of diffuse axonal injury are seen microscopically but the focal lesions in the corpus callosum and brain stem may be seen macroscopically. The typical histological features can be seen two hours after trauma (SCHOLTZ and CHAN 1985) and are characterised in silver preparations by sinusoidal distended axons. At 16 hours tuberosities develop on the axons, and at 24 bullous expansions appear on the torn end of axons – the so-called retraction balls. There is a progressive astrocytic reaction associated with this, best seen following staining with GFAP. Initially foot processes around blood vessels are prominent, followed by easily observed large reactive astrocytes. Over a period of months to years glial scars develop with degeneration of long fiber tracta in the white matter in patients who survive. The clinical correlate of coma is axonal injury (ADAMS et al. 1982).

Diffuse axonal injury was first described in patients with severe post traumatic dementia (STRICH 1956). It is seen at its most marked following severe head injury where there has been no lucid internal. The brain shows grossly enlarged ventricles reflecting the loss of white matter in individuals who survive several years after a head injury (Fig. 14b). The patients typically have a lower incidence of skull fracture, intracranial haematoma and cerebral contusion than patients who have a lucid internal. In addition the incidence of raised intracranial pressure is less in the former group (ADAMS et al. 1982). Where the angular acceleration is less, haematomata are found (GENARELLI et al. 1982). In the most severe cases the disconnection of the cortex leads to coma and the vegetative state (JENNETT and PLUM 1972). Multiple petechial haemorrhages may be seen in association with diffuse axonal injury, these have a similar biomechanical substrate (ADAMS et al. 1986). These authors postulate that large intracerebral haematomata develop from these small haemorrhages.

Hypoxic brain damage is a common accompanyment of head injury (GRAHAM et al. 1978) most noticable in Ammon's horn and the basal ganglia. Associated hypotension will also result in ischaemia and infarction in the boundary zones between the large intracranial arteries. Occasionally thrombosis in a large intracranial vessel will occur and produce extensive infarction.

Intracranial haematoma is an important event following head injury and is an major cause of death in patients who have a lucid internal. Commonly these patients have a skull fracture with a vessel being torn at the time of impact. The haematoma may be extradural, with bleeding from the meningial vessels or subdural where there is bleeding from the bridging veins in the subdural space. The effects produced are those of a space occupying lesion but much more importantly there is associated severe cerebral oedema either unilateral or bilateral. The effects of the raised intracranial pressure are to compress brain structures against the tentorium cerebelli and if the posterior cerebral artery is occluded cerebral infarction will occur.

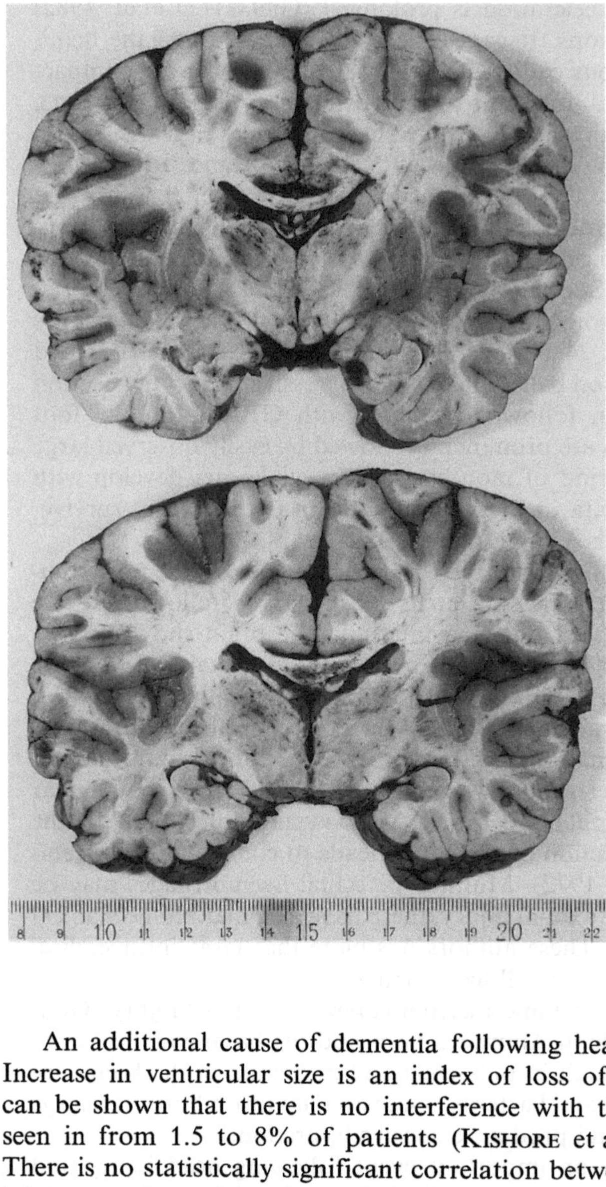

Fig. 14a. Coronal sections of the brain from a patient who died 2 weeks after a severe closed head injury. Note the focal haemorrhages in the corpus callosum

An additional cause of dementia following head injury is hydrocephalus. Increase in ventricular size is an index of loss of brain substance only if it can be shown that there is no interference with the flow of CSF. It can be seen in from 1.5 to 8% of patients (Kishore et al. 1978) with head injury. There is no statistically significant correlation between the severity of the head injury as measured by the period of unconsciousness and the subsequent size of the lateral ventricles (Hawkins et al. 1976; Kishore et al. 1978). However there is an association between increase in size of ventricles and residual disability. In patients with the most severe head injuries there may be an obstructive component of the head injury because of associated subarachnoid haemorrhage (Hawkins et al. 1976). The ventricular enlargement is usually seen by the 14–19th day (Kishore et al. 1978). The acute deterioration of patients after this period with increased enlargement of ventricles suggests a degree of obstruction.

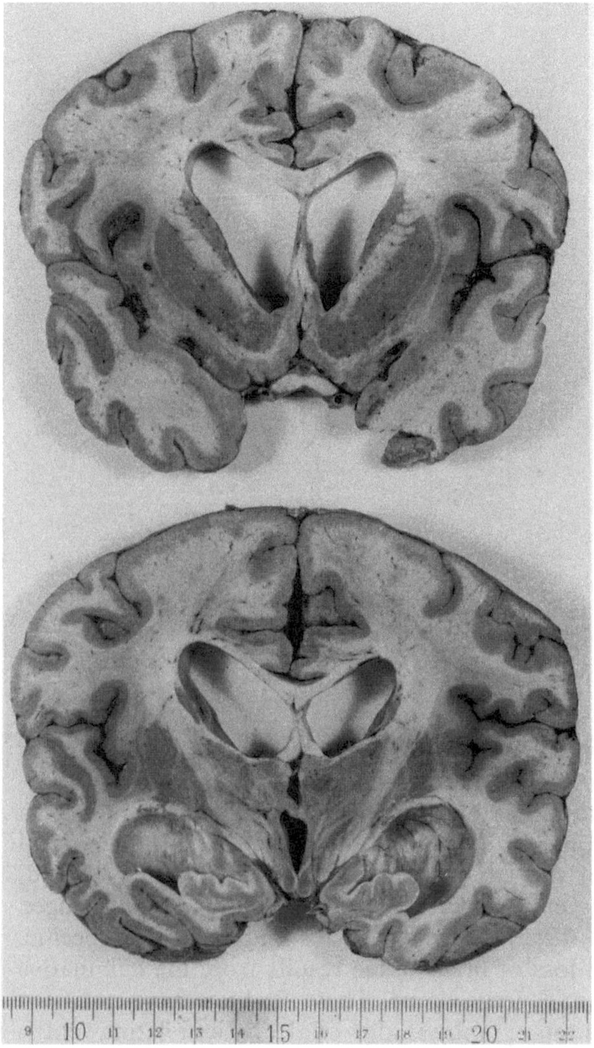

Fig. 14b. Coronal slices of the brain of a patient who died 10 years after a severe closed head injury. Note the enlarged ventricles and thinning of the corpus callosum in the lower slice

Ischaemia can also result from spasm of the cerebral vessels and has been found to occur in between 2–31% of cases with severe head injury (SUWANWELA and SUWANWELA 1972).

9 Alcohol and Dementia

That chronic alcohol ingestion may be associated with a dementing illness "Alcohol dementia" (DREYFUS 1976) is a common observation that has been recorded for over a century. Alcohol may not be the prime cause of brain damage,

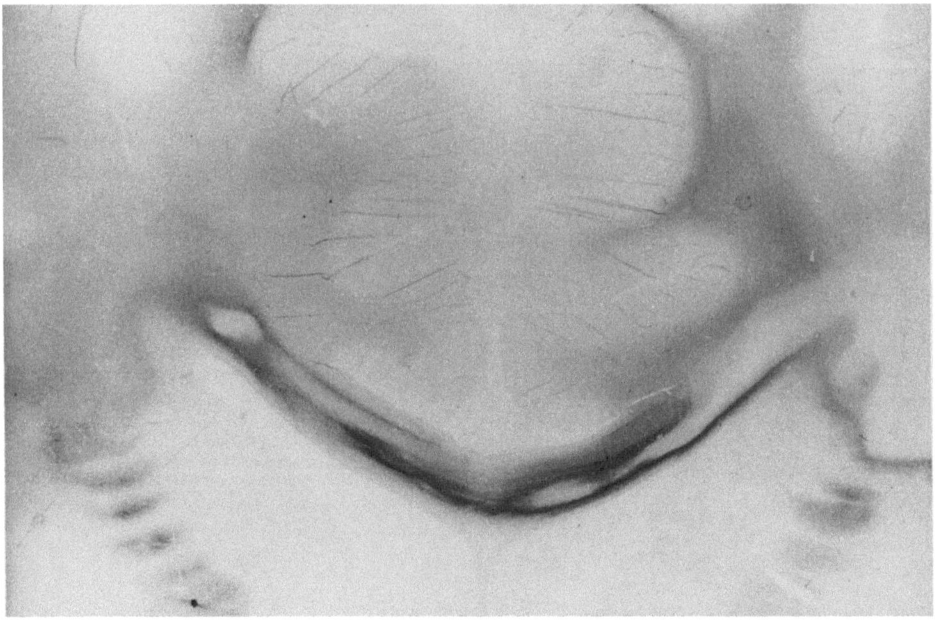

Fig. 15. Low power photograph of section of the corpus callosum in an alcoholic showing necrosis of myelin. Marchi, ×4

the accumulation of fatty acid ethyl esters or acetaldehyde is probably more significant (LAPOSATA and LANGE 1986). The findings in the brain are non-specific (COURVILLE 1955) with slight but significant brain weight reduction (TORVIK et al. 1982) with atrophy of the frontal and parietal lobes, meningeal thickening and enlargement of the lateral ventricles, which may be marked. The observation that enlarged ventricles can be seen in young alcoholics aged 25 to 38 years (TUMARKIM et al. 1955) suggests that this finding may be specific. These authors suggest that the loss of brain tissue results from the cumulative effects of alcohol on neurons and nerve tissue. These findings have not been supported by computer assisted tomography studies of alcoholics relating ventricle size to cognitive loss (CHARNESS and DIAMOND 1984) however more recent neuropathological study (HARPER and KRIL 1985) has shown that there is significant loss of brain tissue which is more marked in alcoholics with superadded Wernicke's encephalopathy. Histological study of the cortex shows astrocytic proliferation and nerve cell loss in the third layer of the frontal cortex (MOREL 1939). Cerebellar ataxia has also been observed in young alcoholics (SKILLICORN 1955). The lesions in the cerebellum tend to be focal in the anterior vermis and the anterior lobes of the cerebellar hemispheres (ALLSOP and TURNER 1966). Microscopical study shows loss of the granular cell layer and Purkinje neurons.

Two more specific lesions seen in alcoholics are secondary to necrosis of myelin; in the corpus callosum (Fig. 15) in the Marchiafava Bignami syndrome (MARCHIAFAVA and BIGNAMI 1903) and pons-central pontine myelinolysis (Fig. 16) (ADAMS et al. 1959). While both lesions can be seen in alcoholics (the

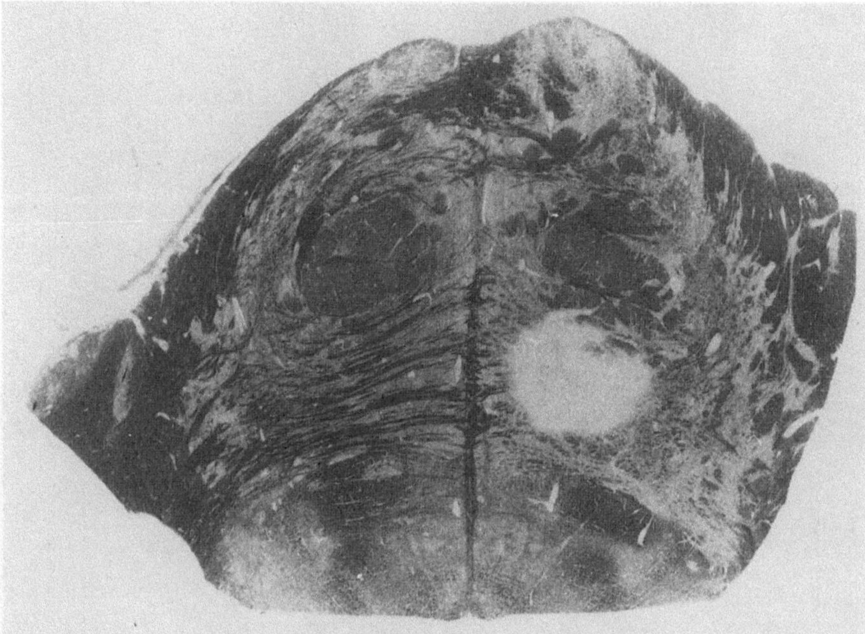

Fig. 16. Necrosis of myelin in pons-central pontine myelinolysis from an alcoholic. Lunol Fast Blue, ×5

former following ingestion of large amounts of cheap red wine) and may be associated with dementia in a subacute form, both can be observed in other clinical settings where patients have had rapid correction of hyponatremia. Central pontine myelinolysis can be produced experimentally following rapid correction of hyponatremia (LAURENO 1982).

The most widely known lesion in chronic alcoholics is that described by WERNICKE (1881) under the term "acute superior haemorrhagic polioencephalitis" characterised by mental disturbance, impairment of memory and disorientation. Pathological changes are located around the third and fourth ventricles, the aqueduct and in the mamillary bodies. Typically, haemorrhages and focal areas of necrosis are visable macroscopically. These features may be seen commonly at autopsy with an incidence of 1.7% of selected coroners autopsies in Perth, Western Australia (HARPER 1979). Histologically, there are dilated blood vessels with prominent endothelial cells surrounded by histiocytes (Fig. 17a, b). In the later stages, in more severe cases, there is extensive tissue necrosis of the mamillary bodies and a prominent macrophage infiltrate. When the mamillary bodies are severely shrunken with neuronal loss and gliosis the patients have gross memory defect initially described by Korsakow (KORSAKOFF 1890) where the memory of recent events is lost primarily. The specificity of the pathology and memory loss has attracted considerable interest. The mamillary bodies may be the only structure involved in the Korsakoff syndrome (BRIERLEY 1966) but more characteristically gliosis and neuronal loss occurs bilaterally in the medial thalamus.

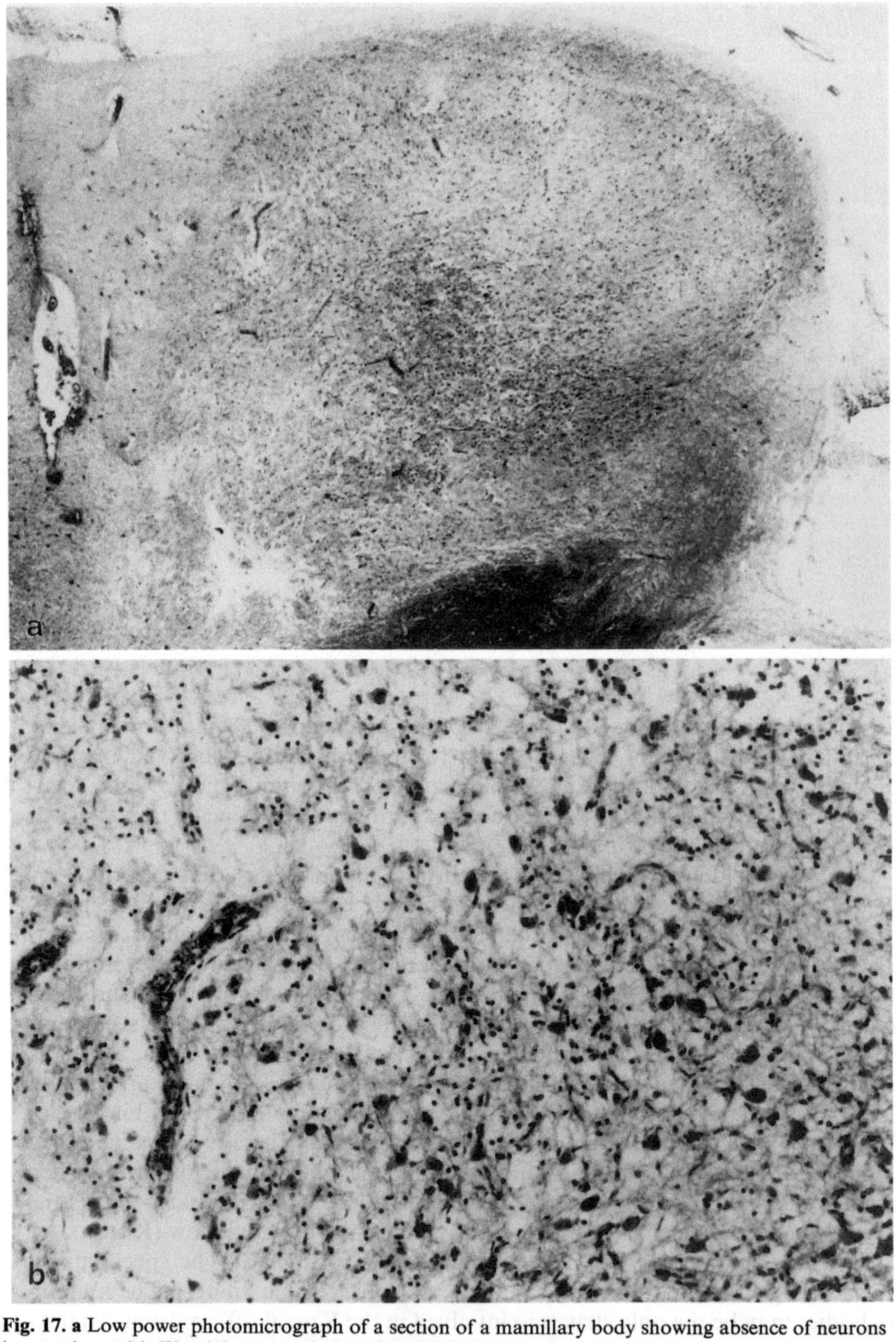

Fig. 17. a Low power photomicrograph of a section of a mamillary body showing absence of neurons in a patient with Wernicke's encephalopathy. H&E, ×7. **b** High power photomicrograph to show prominent blood vessels and absence of neurons and the astrocytic reaction in the mamillary body of a patient with Wernicke's encephalopathy. H&E, ×129

The common association of thiamine deficiency with alcoholism may be secondary to impaired absorption in addition to reduced uptake and metabolism. Thiamine is not synthesized in the central nervous system and must be transported to it. The specifity of the lesion in the periaqueductal tissue and mamillary bodies has been related to deficiency of thiamine on serotoninergic neurons (PLAITAKIS et al. 1982) or the raised pyruvate and lactic acid levels which result from the impaired oxidation of pyruvic acid. In addition the succeptibility of some individuals to Wernicke's syndrome suggests that there may be an underlying genetic predisposition, possibly due to deficiency of the enzyme transketolase which is thiamine dependent (BLASS and GIBSON 1977). An alternative suggestion is that magnesium deficiency, a common observation in alcoholics, interferes with the function of this enzyme (FLINK 1978). Pellagra may be seen in alcoholics characterised by fluctuating confusional states and myoclonic jerks and oppositional hypertonia either on admission to hospital or after receiving treatment (ISHII and NISHIHARA 1981). The histological hallmark is central chromatolysis of neurons particularly in the Betz cells and pontine nuclei.

10 Normal (Intermittently Raised) Pressure Hydrocephalus

The initial description of dementia developing over months, with gait disturbance and ultimately urinary incontinence was made in three patients with "normal" CSF pressure on lumbar puncture (ADAMS et al. 1965). On continuous monitoring of intracranial pressure episodes of moderate intracranial hypertension were observed (PICKARD et al. 1980) which gives this entity the more appropriate name of intermittently raised pressure hydrocephalus. Up to 5% of demented individuals with this clinical presentation may have large ventricles. In many there will be an obvious explanation of the failure to resorb cerebrospinal fluid. Obliteration of the subarachnoid space can result from subarachnoid haemorrhage following severe head injury (HAKIM and ADAMS 1965), ruptured berry aneurysm or following meningeal infection. However there is a group for which there is no explanation, the patients are unable to compensate for infusion of artificial CSF (KATZMAN and HUSSEY 1970) which suggests that the CSF resorptive mechanisms are impaired (SYMON and DORSCH 1975). Once the absorptive facility is damaged and the normal adaptive mechanisms have been exhausted there is a marked transient increase in CSF pressure. In the final stage when the compensatory mechanisms are fully utilised the CSF pressure is continually raised. The importance of this group of patients is that shunting may offer preservation and improvement of intellect (HUGHES et al. 1978).

11 Huntington's Chorea

Huntington's chorea is a serious progressive though rare disorder following autosomal dominant inheritance. One in four cases are the result of new mutations. It is generally clinically manifest between 25–45 years and has a course of 15 years, with death in the 50's. Close genetic linkage has recently been

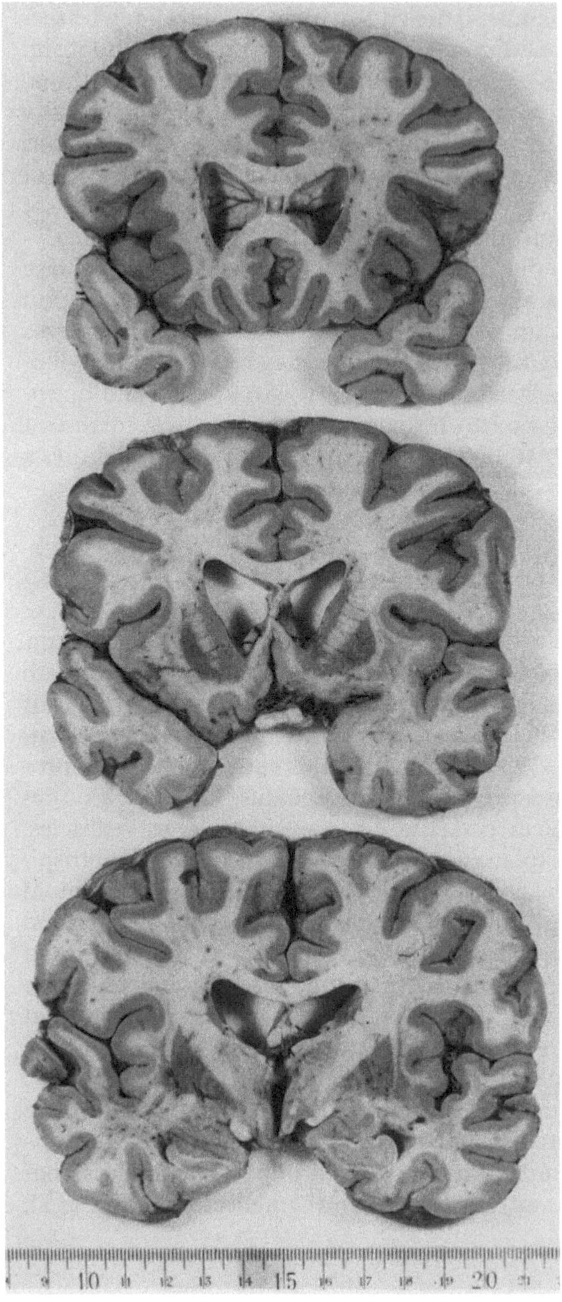

Fig. 18. Coronal slides of the brain of a patient with Huntington's Chorea showing atrophy of the head of the candate nucleus, most noticeable in the lower slice

found between Huntington's chorea and a polymorphic DNA probe G8 on chromosome 4 (GUSELLA et al. 1983). The mode of development of the syndrome is variable including the order of appearance and relative prominence of memory deterioration, personality change and movement disorder. Initially there is a combination of semi-purposful movements with dementia and either may dominate the picture. Where dementia dominates the picture the term Huntington's disease is applied and this pattern may be familial.

On external appearance the brain is commonly normal through there may be uniform cortical thinning with reduction of brain weight to less than 1000 g. The major abnormality seen on slicing the brain is thinning of the corpus striatum which is usually diagnostic (Fig. 18). The head of the caudate nucleus is flattened and does not protrude into the lateral ventricle. The putamen and globus pallidus are commonly also reduced in size. The white matter, particularly in the corpus callosum, can be reduced significantly. The brain stem and cerebellum generally shows minor changes only.

Microscopic examination of the corpus striatum shows selective loss of the small neurons from a normal ratio of small to large neurons of 160:1 to 40:1 (LANGE et al. 1976). There is an apparent increase in the number of astrocytes with gliosis which probably represents packing of the astrocytes as the small cells are lost (LANGE et al. 1976). The neurons in the subthalamic nuclei are also reduced in size with associated neuronal loss. Rarely are Purkinje neurons lost and then usually only in the juvenile form.

The major neurochemical abnormality is reduced gama-aminobutyric acid and glutamic acid decarboxylase in the substantia nigra, putamen and caudate nucleus. Similarly the synthetic enzymes for acetylcholine synthesis are reduced, but not those for dopamine. Somatostatin is increased (BEAL et al. 1984). The neurons in the caudate nucleus that contain the enzyme nicotinamide adenine dinucleotide phosphate diaphorase are preserved, suggesting that this enzyme makes them resistent to the degenerative process (FERRANTE et al. 1985). The changes in the cortex are primarily neuronal loss, most marked in layer three.

The chorea of Huntington's disease results from the loss of these intrinsic and efferent structural neurons and the resultant inability of the corpus striatum to select behaviour to carry out or suppress (MARSDEN 1980). There is loss of the integrity of a positive feed back circuit from the motor cortex through the striatum, medial globus pallidus and thalamus, returning to the cortex (PENNY and YOUNG 1983).

12 Lysosomal Enzyme Deficiences

The lysosomal enzyme deficiencies form a group of disorders where there is a block in the intralysosomal digestion with the accumulation of material in enlarged lysosomes. This observation can be supported by the presence of acid phosphatase in relation to these large vacuolar structures. Several of these disorders can present in adult life with dementia as an important feature. Many of them can be diagnosed by non-invasive tests and brain biopsy is rarely necessary to reach a diagnosis. The most important ones seen in adult life are the Ganglosidoses Gm2 (RAPIN et al. 1976) Gm1 (KOBAYASHI and SUZUKI 1981), neuronal ceroid lipofuscinosis (Kufs disease) (GOEBEL et al. 1982). Niemann Pick disease, Gauchers disease type three and metachromatic leucodystrophy (KRAUS-RUPPERT et al. 1972).

13 Picks Disease

Picks disease is a dementing illness of unknown aetiology occurig in middle age with a peak at 60 but may be seen in the 20's or the elderly. The course of the disease is short from 1 to 10 years, and it is slightly more common in women. There may be a family history. Clinically it presents with intellectual deterioration, difficulty in concentration and loss of memory with progression to severe dementia.

The macroscopic appearance of the brain is lobar atrophy which is relatively circumscribed and may be extreme. It is usually frontotemporal but may select either lobe with a tendancy to involve the left half of the brain more than the right. The gyri appear knife-edge-like with loss of the underlying white matter. When the tissue loss is extreme the brain can weigh less than 1000 g. The loss can involve the caudate nucleus giving an appearance which resembles Huntington's chorea. The microscopic appearance of the cortex is of neuronal loss and gliosis. There is however a degenerate appearance of neurons which is commonly seen in Picks disease. The cell body swells up and becomes round or pear shaped. The cytoplasm is homogenous and pink on haematoxylin and eosin staining or diffusely stained with Mallorys phosphotungstic acid-haematomylin, displacing the nucleus to the side (Fig. 19). This histological feature is seen both on large and small neurons in the cortex, basal ganglia and brain stem. Electronmicroscopy of these cells show that the Picks bodies are accumula-

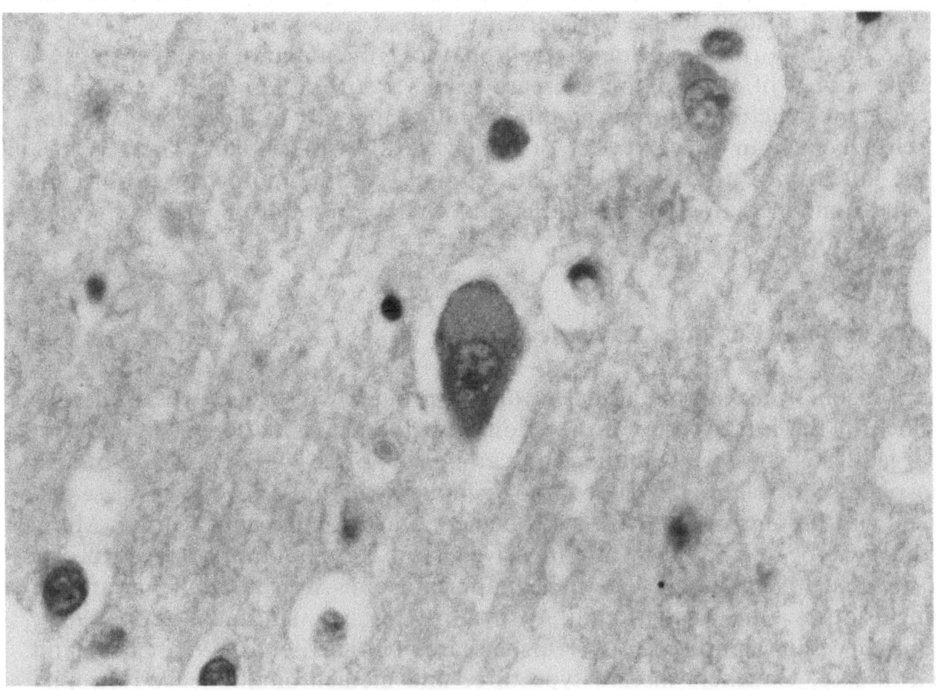

Fig. 19. A photomicrograph of a globoid nuclei from a patient with Pick's disease. H&E, ×174

tions of filaments and tubules which probably represent response of the neuron to damage (WISNIEWSKI et al. 1972). The distribution of the Picks bodies is similar to that in Alzheimer's disease but there is no other association (BALL 1979). The attempt to transmit the disease to laboratory animals has so far been unsuccessful.

14 Tumours and Dementia

Tumours, both primary and secondary, may produce a dementing illness on the central nervous system both by direct or remote effects. The primary brain tumours most commonly associated with dementia are the slowly growing ones, particularly meningiomas when the destroy the frontal lobes or anterior corpus callosum (Fig. 20). Malignant gliomas (Fig. 21) and secondary carcinoma usual-

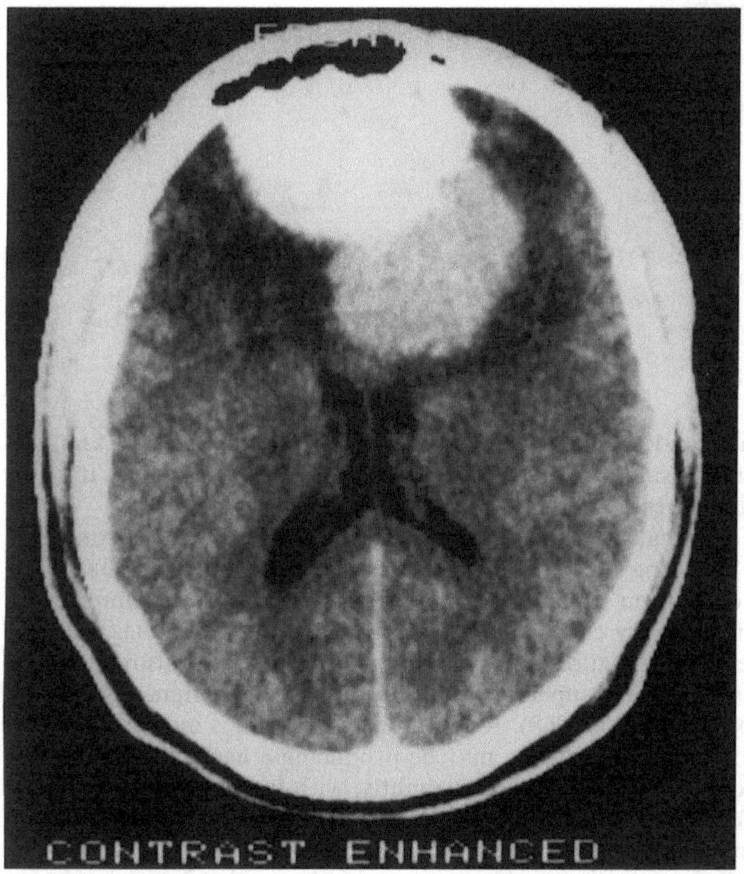

Fig. 20. A brain scan of a 52-year-old man with a years history of progressive dementia showing a bilobed frontal meningioma producing local oedema and posterior displacement of the ventricles. He made full recovery after removal

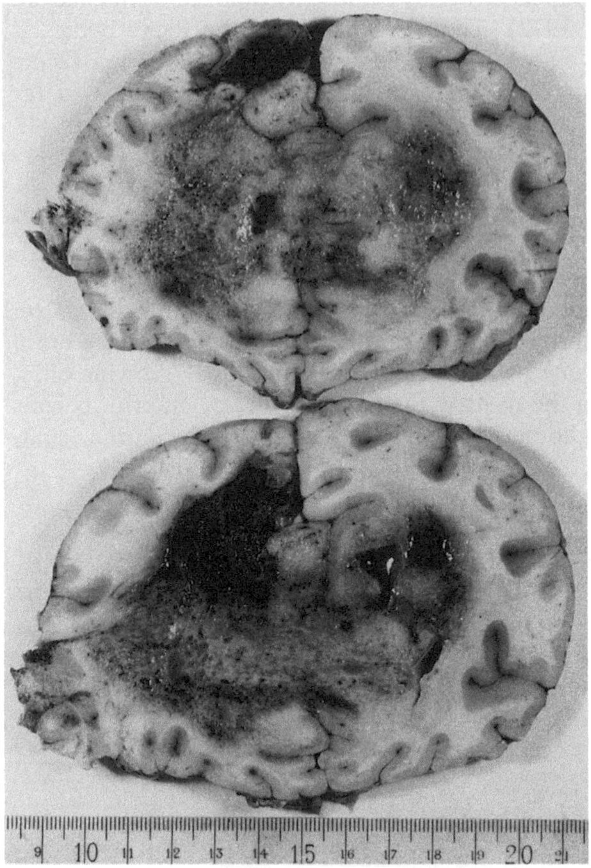

Fig. 21. Coronal slices of a demented patient with a large frontal and callosal glioblastoma

ly run a shorter course and the dementia is a minor part of the illness. Rarely secondary tumour in the subarachnoid space (carcinomatous meningitis) may present as a dementing illness.

The remote effects of malignant neoplasms or the paraneoplastic syndromes effect many organs including the central nervous system (HENSON and URICH 1979). Anxiety and depression occur early and are almost constant symptoms. Disturbance of recent memory can be an outstanding feature (CORSELLIS et al. 1968). Convulsive epileptic seizures are occasionally the initial feature, with hallucinations. Dementia usually supervenes and the underlying malignancy may not be apparent (CORSELLIS et al. 1968). A common association is oat cell carcinoma – a neuroendocrine tumour which may be producing a neurotransmitter-like substance. Histological evidence of encephalitis, with lymphocytes around blood vessels and focally present in the hippocampal formation, amygdaloid nuclei, cingulate and cerebral cortex have been described (HENSON et al. 1965). The histological findings are non-specific however and may be found in the absence of underlying systemic pathology (HENSON and URICH 1982).

15 Dialysis Dementia

Dialysis Dementia is an uncommon complication of long term haemodialysis and may have an insiduous or rapid onset. Speech disorder is early and prominent with myoclonus, ataxia and a progressive dementia which may fluctuate in relation to haemodialysis. The encephalopathy is associated with excessive amounts of aluminium in tap water (ALFREY et al. 1976; ELLIOT and MACDOUGAL 1978) and with high level of aluminium in the serum and brains of these patients (ALFREY et al. 1972; MCDERMOTT et al. 1978; CRAPPER et al. 1978).

No specific neuropathological abnormality has been noted in the brain in this disorder (PARKINSON et al. 1981). Loss of Purkinje neurons was reported (BURKS et al. 1976). Aluminium poisoning in experimental animals and the culture of human neurons exposed to aluminium (CRAPPER et al. 1978) results in the accumulation of neurofilaments in neurons (KLATZO et al. 1965) followed by neuronophagia and astroglial proliferation. One preliminary case report has identified neurofibrillary tangles in dialysis encephalopathy but this has yet to be substantiated (KOGEORGOS and SCHOLTZ 1982). The neurofibrils produced experimentally are unpaired and 100 Å in diameter (KLATZO et al. 1978) unlike those seen in Alzheimer's disease which are paired (KIDD 1964).

16 Liver Failure: Hepatic Encephalopathy

Chronic hepatic failure can be associated with a slow intermittent progressive dementia with superimposed slurred speach, tremor of the hands, cerebellar ataxia, chorioathetosis and electroencephalographic abnormalities. The liver failure may be of any acquired type or due to hereditary Wilsons disease (WILSON 1912). Neurological changes remain in those who survive the hepatic coma.

In acquired liver disease the pathological findings are mainly microscopic and are found at the junction of the cortex and white matter, corpus striatum and cerebellum (VICTOR 1974; FINLAYSON and SUPERVILLE 1981). At these sites there is increase in size and number of protoplasmic astrocytes and a diffuse patchy degeneration of neurons and myelin. Alzheimer's type two astrocytes are evident characterised by large pale lobulated or irregular nuclei with one or two nucleoli (Fig. 22). The nuclei have a distinct nuclear membrane. Where cell degeneration is extensive microcavitation results. In the brain of patients with Wilsons disease the corpus striatum may appear strunken on macroscopic examination and may have a brick-red to brownish colour. There may be cavitation around blood vessels and occasionally in the putamen. Cellular proliferation is easily observed on microscopic examination due to the presence of Alzheimer's type two cells.

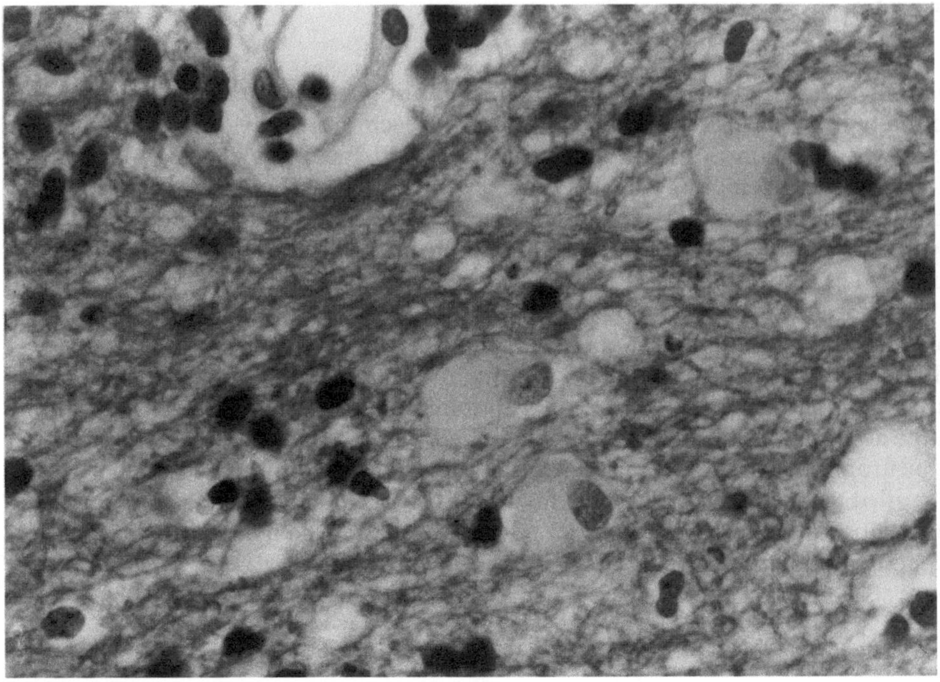

Fig. 22. Photomicrograph of a patient with hepatic encephalopathy to show Alzheimer-type two astrocytes. PTAH, ×130

17 Dementing Disease with Poorly Defined Pathology

Dementia may occur in several clinical situations where the pathology is ill-defined. Endocrine disease, particularly occult thyroid disease, is relatively common in post menopausal women (TUNBRIDGE et al. 1977), and is associated with acute and subacute mental changes (OLIVARIUS and RODER 1970). The pathology described is that of secondary effects of the underlying disease process namely hypotension and anoxia.

Mental changes from neuroasthenia to confusional states and madness (megaloblastic madness) have been described with vitamin B12 deficiency. Pathology has been reported in the white matter (FERRARO et al. 1945) and is similar to that seen in the cord in Subacute Combined degeneration. There is fusiform swelling of myelin sheaths and axons progressing with macrophages and in older lesions gliosis.

The major non-organic psychosis including those which are drug induced can be associated with dementia (drug ingestion or withdrawal, particularly Bromides and Barbiturates) and have no clearly defined specific neuropathology to explain the symptoms.

In schizophraemia (JOHNSTONE et al. 1978) and to a lesser extent in affective psychosis (NASRALLAH et al. 1982) computer assisted tomographic scans show cerebral atrophy with ventricular enlargement. This may represent a drug effect as the brain abnormalities are seen most commonly in patients who have tardive dyskinesia which has an organic component (Famuyiwa et al. 1979). However no histological abnormality either neuronal loss or gliosis has been convincingly demonstrated.

References

Adams JH, Scott G, Parker LS, Graham DI, Doyle D (1980) The contusion cortex: a quantitative approach to cerebral contusions in head injury. Neuropathol Appl Neurobiol 6:319–324

Adams JH, Graham DI, Murray LS, Scott G (1982) Diffuse axonal injury due to non-missile head injury in humans. An analysis of 45 cases. Ann Neurol 12:557–563

Adams JH, Doyle D, Graham DI, Lawrence E, McKlellan DR (1986) Deep intracerebral ("Basal ganglia") haematomas in fatal non-missile head injury in man. J Neurol Neurosurg Psychiatry 49:1039–1043

Adams RD, Victor M, Mancall EC (1959) Central pontine myelinolysis. Arch Neurol Psychiatry 81:154–172

Adams RD, Fisher CM, Hakims S, Ojemann RG, Sirect WH (1965) Symptomatic occult hydrocephalus with "normal" cerebrospinal fluid pressure. A treatable syndrome. N Engl J Med 273:117–126

Alfrey AC, Mishell JM, Burks J (1972) Syndrome of dyspraxia and multifocal seizures with chronic haemodialysis. Trans Am Sec Intern Ong 18:257–261

Alfrey AC, Le Gendre GR, Kachny WD (1976) The dialysis encephalopathy syndrome; possible aluminium intoxication. Engl J Med 294:184–188

Allsop J, Turner B (1966) Cerebellar degeneration associated with chronic alcoholism. J Neurol Sci 3:238–258

Astrosom KE, Mancall EL, Richardson EP (1958) Progressive multifocal encephalopathy. Brain 81:93–111

Ball MJ (1978) Topographic distribution of neurofibrillary tangles and granulovacuolar degeneration in hippocampal cortex of ageing and demented patients – a quantitative study. Acta Neuropathol (Berl) 42:73–80

Ball MJ (1979) Topography of Picks inclusion bodies in hippocampus of demented patients. Acta Neuropathol (Berl) 42:73–80

Beal MF, Bird ED, Langlais PJ, Martin JB (1984) Somatostatin is increased in the nucleus, accumbens in Huntington's disease. Neurology 34:663–666

Beal MF, Mazurek MF, Tran VT, Chattha G, Bird ED, Martin JB (1985) Reduced numbers of Somatostatin receptors in the cerebral cortex in Alzheimer's disease. Science 229:289–291

Behan PO, Behan WMH (1979) Possible immunological factors in Alzheimer's Disease: early recognition of potentially reversible deficits. In: Glen AIM, Walley LJ (eds) Churchill Livingstone, Edinburgh London New York

Benton JS, Bowen DM, Allen SJ, Haan EA, Davison AN, Neary D, Murphy RP, Snowden JS (1982) Alzheimer's disease as a disorder of isodendritic core. Lancet 1:456

Binswanger O (1894) Die Abgrenzung der allgemeinen progressiven Paralyse. Berl Klin Wochenschr 31:1137–1139

Blackwood W, Hallpike JF, Kocen RS, Mair WGP (1969) Atheromatous disease of the carotid arterial system and embolism from the heart in cerebral infarction. A morbid anatomical study. Brain 92:897–910

Blass JB, Gibson GE (1977) Abnormality of a thiamine-requiring enzyme in patients with Wernicke-Korsakoff Syndrome. N Engl J Med 297:1367–1370

Blessed G, Tomlinson BE, Roth M (1968) The association between quantitative measures of dementia and of senile changes in cerebral grey matter of elderly subjects. Br J Psychiatry 114:797–811

Bondareff W, Mountjoy CQ, Roth M (1982) Loss of neurons of the adrenergic projections to cerebral cortex (Nucleus locus ceruleus) in senile dementia. Neurology 32:164–168

Bouteille M, Fontaine C, Vedrenne CL, De la Rue J (1965) Sur un cas d'encéphalite subaigue à inclusions, étude anatomoclinique et ultrastructurale. Rev Neurol 113:454–458

Bowen DM, Smith CB, White P, Flack RHA, Carrasco L, Gedye JL, Davison AN (1977) Chemical pathology of the organic dementias II. Quantitative estimation of cellular changes in post mortem brains. Brain 100:427–453
Brierley JB (1966) The neuropathology of amnesic states. In: Whitty CWM, Nangwill O (eds) Amnesia. Butterworth, London, pp 150–180
Brody H (1955) Organisation of the cerebral cortex III. A study of ageing in the human cerebral cortex. J Comp Neurol 102:511–556
Burks JS, Huddlestone J, Alfrey AC, Noreberg MD, Lewin E (1976) A fatal encephalopathy in chronic haemodialysis patients. Lancet 1:764–768
Candy JM, Oakley AE, Klinowski J, Carpenter TA, Perry RH, Atack JR, Perry EK, Blessed G, Fairbairn A, Edwardson JA (1986) Aluminosilicates and senile plaque formation in Alzheimer's disease. Lancet 1:354–357
Casanova MD, Walker LC, Whitehouse PJ, Price DL (1985) Abnormalities of the nucleus basalis in Down's syndrome. Ann Neurol 18:310–313
Charness ME, Diamond F (1984) Alcohol and the nervous system. In: Affel SH (ed) Current neurology, vol 5. Wiley, New York
Colon EJ (1972) The elderly brain. A quantitative analysis of the cerebral cortex in two cases. Psychiatr Neurol Neurochir (Amst) 75:261–270
Cork LC, Sternberger NH, Sternberger LA, Casanova MF, Struble RG, Price DL (1986) Phosphorylated neurofilament antigens in neurofibrillay tangles in Alzheimer's Disease. J Neuropathol Exp Neurol 45:56–64
Corsellis JAN (1962) Mental illness and the ageing brain. Oxford University Press, London
Corsellis JAN (1976) Some observation on the Purkinje cell population and on brain volume in human ageing. In: Terry RD, Gershaw S (eds) Neurobiology of ageing, vol 3. Raven Press, New York, p 205
Corsellis JAN, Evans PH (1965) The relation of stenosis of the intracerebral arteries to mental disorder and cerebral degeneration in old age. Excerpta Med Int Congr Ser 100:546
Corsellis JAN, Goldberg GJ, Norton AR (1968) "Limbic encephalitis" and its association with carcinoma. Brain 91:481–496
Corsellis JAN, Bruton CJ, Freemen-Browne D (1973) The aftermath of boxing. Psychol Med 3:270–303
Courville C (1955) The effects of alcohol on the nervous system. San Lucas Press, Los Angeles
Coyle AJ, Crow TJ, Johnson JA, Joseph MH, Perry EK, Perry RH, Blessed G, Tomlinson BE (1983) Monoamine metabolism in senile dementia of Alzheimer type. J Neurol Sci 60:383–392
Crapper DR, Krishnan SS, Quittkat S (1976) Aluminium neurofibrillary degeneration and Alzheimer's disease. Brain 99:67–80
Crapper DR, Karlik S, De Boni V (1978) Aluminium and other metals in senile (Alzheimer) dementia. In: Katzman R, Terry RD, Beck KL (eds) Alzheimer's disease: senile dementia and related disorders. Ageing, vol 7. Raven Press, New York, pp 471–485
Creutzfeldt HG (1921) Über eine eigenartige Erkrankung des zentralen Nervensystems. In: Nissl F, Alzheimer A (eds) Histologische und histopathologische Arbeiten über die Großhirnrinde, Suppl 6. Gustav Fischer, Jena, pp 1–48
Cross AJ, Crow TJ, Perry EK, Perry RH, Blessed G, Tomlinson BE (1981) Reduced dopamine, beta-hydroxylase activity in Alzheimer's disease. Br Med J 282:93–94
Cross RB (1982) Demonstration of neurofibrillary tangles in paraffin sections: a quick and simple method using a modification of Palmgren's method. Med Lab Sci 39:67–69
Davies P, Katzman R, Terry RD (1980) Reduced somatostatin like immunoreactivity in cerebral cortex from cases of Alzheimer's disease and Alzheimer senile dementia. Nature 288:279–280
Davis PJM, Wright EA (1977) A new method for measuring cranial cavity and its application to the assessment of cerebral atrophy as autopsy. Neuropathol Appl Neurobiol 3:431–358
Dawson JR Jr (1933) Cellular inclusions in cerebral lesions of lethargic encephalitis. Am J Pathol 9:7–15
Dekaban AS, Sadowsky D (1978) Changes in brain weight during the span of human life: Relation of brain weight to body height and body weight. Ann Neurol 4:345–356
De Reuck J, Crevitis L, De Coster S, Sieben B, Becken VH (1980) Pathogenesis of Binswanger's chronic progressive subcortical encephalopathy. Neurology 3:920–928
Dicks D, Myers RE, Kling A (1969) Uncus and amygdala lesions: Effects on social behaviour in the free ranging rhesus monkey. Science 165:69–71

Dom R, Malfroid M, Baro F (1976) Neuropathology of Huntington's chorea. Studies of the vertebro-basal complex of the thalmus. Neurology (Minneap) 26:64–68

Dreyfus PM (1976) Amblyopia and other neurological disorders associated with chronic alcoholism. In: Vinken PJ, Bruyn GW (eds) Handbook of clinical neurology, vol 28. North Holland, Amsterdam, pp 331–347

Duyckaerts C, Hauw JJ, Piette F, Rainsard C, Poulain R, Bethaux P, Escourolle R (1985) Cortical atrophy in senile dementia of Alzheimer-type is mainly due to a decrease in cortical length. Acta Neuropathol (Berl) 66:72–74

Elliot HL, MacDougal AI (1978) Aluminium studies in dialysis encephalopathy. Proc Eur Dial Transplant Assoc 15:157–163

Famuyiwa OO, Eccleston D, Donaldson AA, Garside RF (1979) Tarditive dyskinesia and dementia. Br J Psychiatry 135:500–504

Ferraro A, Arieti S, English WH (1945) Cerebral changes in the course of pernicious anaemia and their relationship to psychic symptoms. J Neuropathol Exp Neurol 4:217–239

Ferrante RJ, Kowall NW, Beal MF, Richardson EP, Bird ED, Martin JB (1985) Selective sparing of a class of striatal neurons in Huntington's disease. Science 230:561–563

Finlayson MH, Superville B (1981) Distribution of cerebral lesions in acquired hepatocerebral degeneration. Brain 10:79–95

Fisher CM (1965) Lacunes: small deep cerebral infarcts. Neurology 15:774–784

Fisher CM (1969) The anterial lesions underlying lacunes. Acta Neuropathol 12:1–15

Flink EB (1978) Role of magnesium depletion in Wernicke-Korsakoff Syndrome. N Engl J Med 298:743–744

Fuller S (1911) A study of the miliary plaques found in brains of the aged. Ann Insanity 68:147–217

Gennarelli TA, Thibault LE, Adams JH, Graham DI, Thompson CJ, Marcincin RD (1982) Diffuse axonal injury and traumatic coma in the primate. Ann Neurol 12:564–574

Glenner GG, Henry JH, Fujilara S (1981) Congophilic angiopathy in the pathogenesis of Alzheimer's degeneration. Ann Pathol (Paris) 1:120–129

Goebel HH, Braak H, Seidel D, Doshi R, Marsden CD, Gulotta F (1982) Morphologic studies on adult neuronal ceroid lipofuscinoss. Clin Neuropathol 1:157–162

Graham DI, Adams JH, Doyle D (1978) Ischaemic brain damage in fatal non-missile head injuries. J Neurol Sci 39:213–234

Greenfield JG (1950) Encephalitis and encephalomyelitis in England and Wales during the last decade. Brain 73:141–166

Gusella JF, Wexler NS, Conneally PM et al. (1983) A polymorphic DNA marker genetically linked to Huntington's disease. Nature 306:234–238

Hachinski VC, Lassen NA, Marshall J (1974) Multi-infarct dementia. A cause of mental deterioration in the elderly. Lancet 2:207–210

Hakim S, Adams RD (1965) The special problem of symptomatic hydrocephalus with normal cerebrospinal fluid pressure. Observations of cerebrospinal fluid hydrodynamics. J Neurol Sci 2:307–327

Harper C (1979) Wernicke's encephalopathy; a more common disease than realised. J Neurol Neurosurg Psychiatry 42:226–231

Harper C, Krill J (1985) Brain atrophy in chronic alcoholic patients: a quantitative pathological study. J Neurol Neurosurg Psychiatry 48:211–217

Harrison PJ (1986) Pathogenesis of Alzheimer's disease beyond the cholinergic hypothesis: discussion paper. J R Soc Med 79:347–352

Hawkins TD, Lloyd AD, Fletcher GIE, Hanka A (1976) Ventricular size following head injury: a clinico-radiological study. Clin Radiol 27:279–289

Henson RA, Urich H (1979) Remote effects of malignant disease: certain intracranial disorders. In: Vinken PJ, Bruyn GW (eds) Handbook of clinical neurology, vol 38. North Holland, Amsterdam

Henson RA, Urich H (1982) Cancer and the nervous system. Blackwell Scientific, Oxford

Henson RA, Hoffman HL, Urich H (1965) Encephalomyelitis with carcinoma. Brain 88:449–464

Herzog AG, Kemper TL (1980) Amygdaloid changes in ageing and dementia. Arch Neurol 37:625–629

Hirano A (1965) Slow, latent and temporate virus infections. In: Gajdusek DC, Gibbs CJ (eds) Monograph, vol 2. National Institute of Health, pp 23–27

Holbourn AHS (1943) Mechanics of head injuries. Lancet 2:438–441

Holbourn AHS (1945) Mechanics of brain injuries. Br Med Bull 3:147–149

Hubbard BM, Anderson JM (1985) Age-related variations in the neuron content of the cerebral cortex in senile dementia of Alzheimer-type. Neuropathol Appl Neurobiol 11:369–382
Hughes CP, Siegel BA, Coxe WS, Grado MH, Grubb RL, Coleman RE, Berg L (1978) Adult idiopathic communication hydrocephalus with and without shunting. J Neurol Neurosurg Psychiatry 29:154–156
Iqbal K, Wisniewski HM, Shelanski ML, Brostoff S, Liwncz BH, Terry RD (1974) Protein changes in senile dementia. Brain Res 77:337–343
Iqbal K, Wisniewski HM, Grundke-Iqbal I, Korthals JK, Terry RD (1975) Chemical pathology of neurofibrils. Neurofibrillary tangles of Alzheimer's presenile – senile dementia. J Histochem Cytochem 23:563–569
Iqbal GI, Iqbal K, Wisniewski HM (1982a) Immunocytochemical studies on neurofibrillary changes. In: Hoyer S (ed) The aging brain. Physiological and pathophysiological aspects. Springer, Berlin Heidelberg New York
Iqbal K, Iqbal GI, Merz PA, Wisniewski HM (1982b) Alzheimer neurofibrillary tangles: morphology and biochemistry. In: Hoyer S (ed) The aging brain. Physiological and pathophysiological aspects. Springer, Berlin Heidelberg New York
Ishii N, Nishihara Y (1981) Pellagra among chronic alcoholics: clinical and pathological study of 20 autopsy cases. J Neurol Neurosurg Psychiatry 44:209–215
Ishii T, Haga S, Tobutakes (1979) Presence of neurofilament protein in Alzheimer's neurofibrillary tangles ANF; an immunofluorescent study. Acta Neuropathol 48:105–112
Jakob A (1921) Über eigenartige Erkrankungen des Zentralnervensystems mit bemerkenswerten anatomischen Befunden mit disseminierten Degenerationsherden. Z Ges Neurol Psychiatrie 64:147–128
Janota I (1981) Dementia, deep white matter damage and hypertension: Binswanger's disease. Psychol Med 11:39–48
Jennett B, Plum F (1972) Persistent vegetative state after brain damage. Lancet 1:734–737
Johnstone EC, Crow TJ, Firth CD, Carney M, Price JS (1978) Mechanism of the antipsychotic effect in the treatment of acute schizophraenia. Lancet 1:848–851
Katzman R, Hussey F (1970) A simple constant infusion manometric test for measurement of CSF absorption. Rationale Method Neurology (Minneap) 20:534–544
Kelly R (1981) The post traumatic syndrome. J R Soc Med 74:242–245
Kidd M (1964) Alzheimer's disease – an electronmicroscopic study. Brain 87:307–319
Kim RC, Collins GH, Paris JE, Wright AW, Chu YB (1981) Familial dementia of adult onset with pathological findings of a 'non-specific' nature. Brain 104:61–78
Kishore PRS, Lipper MH, Girevendulis DP, Becker DP, Vines FS (1978) Post-traumatic hydrocephalus in patients with severe head injury. Neuroradiology 16:261–265
Klatzo I, Wisneiwski H, Streicher E (1965) Experimental production of neurofibrillary degeneration. Light microscopic observations. J Neuropathol Exp Neurol 24:187–199
Kobayashi T, Suzuki K (1981) Chronic Gm1 – gangliosidosis presenting as dystronia 11. Biochemical studies. Ann Neurol 9:476–483
Kogeorgos J, Scholtz CL (1982) Neurofibrillary tangles in aluminium encephalopathy: a new finding. J Neuropathol Appl Neurobiol 8:246
Korsakow SS (1890) Über eine besondere Form psychischer Störung kombiniert mit multipler Neuritis. Arch Psychiatr Nervenkr 21:669–709
Kraus-Ruppert R, Windbolz A, Matthieu JM, Herschkowitz N (1972) The late form of metachromatic leukodystrophy. J Neurol Sci 17:371–381
Ladurner G, Iliff LS, Lechner H (1982a) Clinical factors associated with vascular dementia in ischaemic stroke. J Neurol Neurosurg Psychiatry 45:97–101
Ladurner G, Iliff LD, Sager WD, Lechner H (1982b) A clinical approach to vascular (multi-infarct) dementia. In: Hoyer S (ed) The aging brain. Physiological and pathophysiological aspects. Experimental Brain Research, Suppl 5. Springer, Berlin Heidelberg New York
Lange H, Thorner G, Hope A, Schroder KF (1976) Morphometric studies of the neuropathological changes in choreatic diseases. J Neurol Sci 28:401–425
Laposata EA, Lange LE (1986) Presence of monoridature ethanol metabolism in human organs commonly damaged by ethanol abuse. Science 231:497–499
Larsson T, Sjogren T, Jacobson G (1963) Senile dementia. Acta Psychiatr Scand 39 (Suppl 167): 1–227

Laureno RI (1982) Central pontine myelinolysis following rapid correction of hyponatremia. Ann Neurol 13:232–242
Lewin W, Marshall TF, Roberts AH (1979) Long term outcome after severe head injury. Br Med J 1533–1538
Lhermitte F, Gautier JE, Derouesne C (1970) Nature of occlusions of middle cerebral artery. Neurology 20:82–88
Lishman WA (1978) Organic psychiatry. The psychological consequences of cerebral disorder. Blackwell Scientific, Oxford London Edinburgh Melbourne
Lwoff A (1960) Tumour viruses and the cancer problem; a summation of the conference. Cancer Res 20:820–829
Malone M, Prior P, Scholtz CL (1981) Brain damage after cardiopulmonary bypass: correlation between neuropsychological and neuropathological findings. J Neurol Neurosurg Psychiatry 44:924–931
Mandybur TI (1975) The incidence of cerebral amyloid angiopathy in Alzheimer's disease. Neurology (Minneap) 25(2):120–126
Mandybur TI (1986) Cerebral amyloid angiopathy: The vascular pathology and complications. J Neuropathol Exp Neurol 45:79–80
Mann DMA, Sinclair KGA (1978) The quantitative assessment of lipofuscin pigment, cytoplasmic RNA and nucleolar volume in senile dementia. Neuropathol Appl Neurobiol 4:129–135
Mann DMA, Yates PO, Marcyniuk B (1984) Alzheimer's presenile dementia, senile dementia of Alzheimer-type and Down's syndrome in middle age form an age related continuum of pathological changes. Neuropathol Appl Neurol 10:185–207
Marchiafava E, Bignami A (1903) Sopra un alterazione del corpo calloso osservata in sogetti alcoolista. Riv Patol Nerv Ment 8:544–549
Marsden CD (1980) The enigma of the basal ganglia and movement. Trends Neurosci 3:284–287
Marsden ED, Harrison MFG (1972) Outcome of investigations of patients with presenile dementia. Br Med J 2:249
Masters C, Gajdusek DE, Gibbs EJ (1981a) Creutzfeldt-Jacob disease virus isolation from the Gerstmann-Straussler Syndrome with an analysis of the various forms of amyloid plaque deposition in the virus induced spongiform encephalopathies. Brain 104:535–589
Masters C, Gajdusek DC, Gibbs CJ (1981b) The familial occurence of Creutzfeldt-Jacob disease and Alzheimer's disease. Brain 104:535–558
McDermott JR, Smith AI, Ward MK, Parkinson IS, Kerr NS (1978) Brain aluminium concentration in dialysis encephalopathy. Lancet 1:901–904
McKinley MP, Bolton DC, Prusiner SB (1983) A protease-resistant protein is a structural component of the scrapie prion. Cell 35:57–62
Mitchell DE, Adams JH (1973) Primary focal impact damage to the brain stem in blunt head injuries – does it exist? Lancet 2:215–218
Morel F (1939) Une forme anatomo-clinique particulaire de l'alcoholisme chronique: sclérose corticale laminaire alcoholique. Rev Neurol 71:280–288
Morrison JH, Rogers J, Scherr S, Benoit R, Bloom FE (1985) Somatostatin immunoreactivity in neuritic plaques of Alzheimer's patients. Nature 314:90–92
Nasrallah HA, McCalley-Whitters M, Jacoby CG (1982) Cerebral ventricular enlargement in young manic males. J Affect Disord 4:15–19
Nauta WJH (1961) Fiber degeneration following lesions of the amygdaloid complex in the monkey. J Anat 95:515–531
O'Brien L, Shelley K, Towfighi J, McPherson A (1980) Crystalline ribosomes are present in brains from senile humans. Proc Natl Acad Sci USA 77:2260–2264
Ogata J, Fujishima M, Tamaki K, Nakotomi Y, Ishituska T, Omae T (1980) Stroke-prone spontaneously hypertensive rats as an experimental model of malignant hypertension, light and electron microscopic study of the brain. Acta Neuropathol (Berl) 51:179–184
Olivarius B, Roder E (1970) Reversible psychosis and dementia in myxoedema. Acta Psychiatr Scand 46:1–13
Padgett BL, Walker DL, Zurhein G, Eckroade RJ (1971) Cultivation of papova-like virus from human brain with progressive multifocal leucoencephalopathy. Lancet 1:1257–1260
Parkinson IS, Ward MK, Kerr DNS (1981) Dialysis encephalopathy bone disease and anaemia: the aluminium intoxication syndrome during regular haemodialysis. J Clin Pathol 34:1285–1294

Pearson RC, Esiri MM, Hiorns RW, Wilcock GK, Powell TB (1985) Anatomical correlates of the distribution of the pathological changes in the neocortex in Alzheimer's disease. Proc Natl Acad Sci USA 82:4531–4534

Penny JB, Young AB (1983) Speculations on the functional anatomy of basal ganglia disorders. Ann Rev Neurosci 6:73–94

Perry EK, Tomlinson BE, Blessed G, Bergmann K, Gibson PH, Perry RH (1978) Correlation of cholinergic abnormalities with senile plaques and mental test scores in senile dementia. Br Med J 2:1457–1459

Perry RH, Candy JM, Perry EK, Irving D, Blessed G, Fairbairn A, Tomlinson BE (1982) Extensive loss of choline acetyltransferase activity is not reflected by neuronal loss in the nucleus of Meynert in Alzheimer's disease. Neurosci Lett 33:311–315

Perry RH, Tomlinson BE, Candy JM, Blessed G, Foster JF, Bloxham CA, Perry EK (1983) Cortical cholinergic deposit in mentally impaired Parkinsonism patients. Lancet 2:789–790

Pickard JD, Teasdale G, Matheson M, Lindsay K, Galbraith S, Wyser D, MacPherson P (1980) Intraventricular pressure waves – the best predictive test for shunting in normal pressure hydrocephalus. In: Shulman K, Marmarou A, Miller JD, Becker DP, Hochwald GM, Brock M (eds) Intracranial pressure IV. Springer, Berlin Heidelberg New York, pp 498–500

Plaitakis A, Hwang EC, Van Woert MH, Szilagyi PIA, Berl S (1982) Effect of thiamin deficiency on brain transmitter systems. Ann NY Acad Sci 3701:367–380

Powers JM, Spicer SS (1977) Histochemical similarity of senile plaque amyloid to apudamyloid. Virchows Arch (Pathol Anat) 367:107–115

Prusiner SB, McKinley MP, Bowman KA (1983) Scrapie prions aggregate for form amyloid like birefringent rods. Cell 35:349–358

Rapin I, Suzuki K, Suzuki K, Valsamin MP (1976) Adult (chronic) Gm2 – gangliosidosis. Atypical spinocerebellar degeneration in a jewish sibship. Arch Neurol 33:120–130

Report of the Royal College of Physicians by the College Committee on Geriatrics (1981) Organic mental impairment in the elderly. Implications for research education and the provision of services. J R Coll Physicians Lond 15:141–167

Ridley RM, Baker HF, Crow TJ (1985) Virogenes in scrapie and Creutzfeldt-Jacob disease. J R Soc Med 78:970

Roberts GW, Crow TJ, Polak JM (1985) Location of neuronal tangles in somatostatin neurons in Alzheimer's disease. Science 314:92–94

Rossor MN, Garrett NJ, Johnson AL, Mountjoy CQ, Roth M, Iversen CL (1982) A post-mortem study of the cholinergic and GABA systems in senile dementia. Brain 105:313–330

Scheibel ME, Scheibel AB (1975) Structural changes in the ageing brain. In: Brody H, Harmand D, Ordy JM (eds) Ageing 1. Clinical, morphologic and neurochemical aspects in the ageing central nervous system. Raven Press, New York, pp 11–37

Scholtz CL, Brown A (1978) Lipofuscin and transsynaptic degeneration. Virchows Arch (Pathol Anat) 281:35–40

Scholtz CL, Chan KK (1985) White matter damage following acute head injury. J Neuropathol Appl Neurobiol 11:319

Scholz W (1938) Studien zur Pathologie der Hirngefäße. Die drüsige Entartung der Hirnarterien und Capillaren. Z Ges Neurol Psychiatrie 162:694–715

Schultman S (1956) Bilateral symmetrical degeneration of the thalamus. J Neuropathol Exp Neurol 15:208–209

Sharer LR, Eun-sook C, Epstein LG (1985) Multinucleated giant cells and HTLV-111 in AIDS encephalopathy. Hum Pathol 16:760

Shaw GM, Harper ML, Hahn BH (1985) HTLV-III infections in brains of children and adults with AIDS encephalopathy. Science 227:177–182

Shelanski ML, Wisniewski HM (1969) Neurofibrillary degeneration induced by vincristine therapy. Arch Neurol 20:199–206

Siegal FP (1985) International conference on AIDS. US Dept of Human Services, Atlanta CDC/WHO

Simchowicz T (1924) Sur le signification des plaques seniles et sur la formule de l'écores cerébrals. Rev Neurol 1:221–227

Skillicorn SA (1955) Presenile cerebellar ataxia in chronic alcoholics. Neurology 5:527–534

Smith E (1974) Influence on site of impact on cognitive impairment persisting long after severe closed head injury. J Neurol Neurosurg Psychiatry 37:719
Sofroniew MV, Pearson RCA, Eckenstein F, Evello AC, Powell TPS (1983) Retrograde changes in cholinergic neurons in the basal forebrain of the rat following cortical damage. Brain Res 289:370–374
Strich SJ (1956) Diffuse degeneration of the cerebral white matter in severe dementia following head injury. J Neurol Neurosurg Psychiatry 19:163–185
Suwanwela C, Suwanwela N (1972) Intracranial arterial narrowing and spasm in acute head injury. J Neurosurg 36:314–323
Symon L, Dorsch NWC (1975) Use of long term intracranial pressure measurements to assess hydrocephalic patients prior to shunting. J Neurosurg 42:258–273
Terry RD (1980) Structural changes in senile dementia of the Alzheimer-type. In: Amaducci L, Davison AN, Antuono P (eds) Ageing of the brain and dementia, Ageing, vol 13. Raven Press, New York, pp 23–32
Terry RD, Wisniewski KIH (1970) The ultrastructure of the neurofibrillary tangle and the senile plaque. In: Wastenholme GEW, O'Connor M (eds) Alzheimer's disease and related conditions. A Ciba Foundation Symposium. Churchill, London, pp 145–168
Terry RD, Peck A, De Teresa R, Schechter R, Hordupion DS (1981) Some morphometric aspects of the brain in senile dementia of the Alzheimer-type. Ann Neurol 10:184–192
Tomlinson BE, Corsellis (1984) Ageing and the dementias. In: Adams JH, Corsellis JAN, Duchen LW (eds) Greenfield's neuropathology, vol 4. Edward Arnold, London, pp 977–978
Tomlinson BE, Henderson G (1976) Some quantitative cerebral findings in normal and demented old people. In: Terry RD, Gershon S (eds) Neurobiology of ageing, vol 3. New York, Raven Press, pp 183–209
Tomlinson BE, Blessed GP, Roth M (1970) Observations on the brains of demented old people. J Neurol Sci 11:205–242
Tomlinson BE, Irving D, Blessed G (1981) Cell loss in the locus coerulus in senile dementia of Alzheimer-type. J Neurol Sci 49:419–428
Tomlinson BE, Blessed GP, Roth M (1982) Observations on the brains of non-demented old people. J Neurol Sci 7:331–356
Torack RM (1978) Current evaluation of pathological correlates of dementia. In: The pathologic physiology of dementia. Springer, Berlin Heidelberg New York, pp 77–100
Torvik A, Lindboe CF, Rogde S (1982) Brain lesions in alcoholics. A neuropathologic study with clinical correlations. J Neurol Sci 56:233–248
Tumarkin B, Wilson JD, Snyder G (1955) Cerebral atrophy due to alcoholism in young adults. US Armed Forces Med J 6:67–74
Tunbridge WM, Evered DC, Hall R, Appleton D, Brewis M, Clark F, Evans JG, Young E, Bird T, Smith PA (1977) The spectrum of thyroid disease in a community: the Wickham survey. Clin Endocrinol 7:481–493
Victor M (1974) Neurologic changes in liver disease. Brain dysfunction in metabolic disorders. Res Publ Assoc Res Nerv Ment Dis 53:1–11
Weisenberg RC, Timasheff SM (1970) Aggregation of microtubule subunit protein. Effects of divalent cations colchicine and vinblastine. Biochemistry 9:4110–4116
Weller RO, Steart PV, Powel-Jackson JD (1986) Pathology of Creutzfeldt-Jakob disease associated with pituitary-derived human growth hormone administration. Neuropathol Appl Neurobiol 12:117–129
Wernicke C (1881–1883) Lehrbuch der Gehirnkrankheiten 2. Fischer, Kassel Berlin
Whitehouse PJ, Hedreen JC, White CH, Price DL (1983) Basal forebrain neurons in dementia of Parkinson's disease. Ann Neurol 13:243–248
Wilcock GK, Esiri MM (1982) Plaques, tangles and dementia, a quantitative study. J Neurol Sci 56:343–356
Wilkins RH, Brody IA (1969) Alzheimer's disease (Translation of: Über eine eigenartige Erkrankung der Hirnrinde. Centr Nervenh Psychiat 30:177–179). Arch Neurol 21:109–110
Wilson SAK (1912) Progressive centricular degeneration: a familial nervous disease associated with cirrhosis of the liver. Brain 34:295–509

Wisniewski H, Terry RD (1968) An experimental approach to the morphogenesis of neurofibrillary degeneration and the argyrophilic plaque. In: Wolstenholme GEW, O'Connor M (eds) Alzheimer's disease and related conditions. Churchill, London, pp 223–241
Wisniewski H, Shelanski ML, Terry RD (1964) Effects of mitotic spindle inhibitors on neurotubules and neurofilaments in anterior horn cells. J Cell Biol 39:224–229
Wisniewski HM, Coblenz JM, Terry RD (1972) Picks Disease: a clinical and ultrastructural study. Arch Neurol 26:97–102
Wisniewski HM, Bruce ME, Fraser H (1975) Infectious aetiology of neurotic (senile) plaques in mice. Science 190:1108–1110
Wisniewski HM, Narang HK, Terry RD (1976) Neurofibrillary tangles of paired helical filaments. J Neurol Sci 27:173–181
Wolstenholme GEW, O'Connor M (eds) (1970) Alzheimer's disease and related conditions. Ciba Symposium. Churchill, London
Yamamoto T, Hirano A (1986) A comparative study of modified Bielschowsky, Bodian and Thioflarin S stains of Alzheimer's neurofibrillary tangles. Neuropathol Appl Neurobiol 12:3–9
Zu Rhein GM, Chou SM (1965) Particles resembling papovaviruses in human cerebral demyelinating disease. Science 148:1477–1479

Subject Index

The numbers set in *italics* refer to those pages on which the respective catch-word is discussed in detail

Index of Volumes 72–75 and 77 Current Topics in Pathology

Volume 75: The Human Thymus. Edited by H.K. MÜLLER-HERMELINK

B. VON GAUDECKER, The Development of the Human Thymus Microenvironment

G.G. STEINMANN, Changes in the Human Thymus During Aging

G. JANOSSY, M. BOFILL, L.K. TREJDOSIEWICZ, H.N.A. WILLCOX, M. CHILOSI, Cellular Differentiation of Lymphoid Subpopulations and Their Microenvironments in the Human Thymus

N. DOUROV, Thymic Atrophy and Immune Deficiency in Malnutrition

CH. NEZELOF, Pathology of the Thymus in Immunodeficiency States

H. WEKERLE, H.K. MÜLLER-HERMELINK, The Thymus in Myasthenia Gravis

H.K. MÜLLER-HERMELINK, M. MARINO, G. PALESTRO, Pathology of Thymic Epithelial Tumors

Volume 77: Morphological Tumor Markers. Edited by G. SEIFERT

G.L. WRIGHT JR., A.D. COX, Monoclonal Antibodies to Human Tumor Antigens

T.M. CHU, Immunohistochemical Markers for Human Cancer

H. DENK, Immunohistochemical Methods for the Demonstration of Tumor Markers

R. MOLL, Epithelial Tumor Markers: Cytokeratins and Tissue Polypeptide Antigen (TPA)

G. KLÖPPEL, J. CASELITZ, Epithelial Tumor Markers: Oncofetal Antigens (Carcinoembryonic Antigen, Alpha Fetoprotein) and Epithelial Membrane Antigen

T.O. ROGNUM, P.S. THRANE, F.R. KORSRUD, P. BRANDTZAEG, Epithelial Tumor Markers: Special Markers of Glandular Differentiation

W. ALTMANNSBERGER, M. OSBORN, Mesenchymal Tumor Markers: Intermediate Filaments

H.F. OTTO, R. BERNDT, K. SCHWECHHEIMER, P. MÖLLER, Mesenchymal Tumor Markers: Special Proteins and Enzymes

R.V. IOZZO, Proteoglycans and the Intercellular Tumor Matrix

J. CASELITZ, Basal Membrane Antigen as Tumor Markers

J. CASELITZ, Lectins and Blood Group Substances as "Tumor Markers"

PH.U. HEITZ, Neuroendocrine Tumor Markers

P. KLEIHUES, M. KIESSLING, R.C. JANZER, Morphological Markers in Neuro-Oncology

TH. LÖNING, K. MILDE, Viral Tumor Markers

I. DAMJANOV, Cell, Tissue and Organ Specific Tumor Markers: An Overview

MIX
Papier aus verantwortungsvollen Quellen
Paper from responsible sources
FSC® C105338

If you have any concerns about our products,
you can contact us on
ProductSafety@springernature.com

In case Publisher is established outside the EU,
the EU authorized representative is:
Springer Nature Customer Service Center GmbH
Europaplatz 3, 69115 Heidelberg, Germany

Printed by Libri Plureos GmbH
in Hamburg, Germany